DUCT SYSTEMS
DESIGN GUIDE

This publication was supported by ASHRAE Research Project 1180
under the auspices of ASHRAE Technical Committee 5.2, Duct Design.

Contributors

Jeff Boldt, PE, HBDP
Fellow ASHRAE
Managing Principal—Director of Innovation and Quality
IMEG Corporation
Madison, WI

Pat Brooks, PE
Member ASHRAE
Senior Project Manager
SMACNA
Cincinnati, OH

Kevin Gebke
Member ASHRAE
New Product Development Engineer
DuctSox Corporation
Dubuque, IA

Stephen Idem, PhD
Member ASHRAE
Professor, Department of Mechanical Engineering
Tennessee Tech University
Cookeville, TN

Chris A. Van Rite
Associate Member ASHRAE
Vice President, Sales
MiTek USA
Chesterfield, MO

Bob Reid
Member ASHRAE
Business Development & Engineering
Spiral Pipe of Texas
Houston, TX

Larry Smith, PE
Member ASHRAE
Engineering Manager
Linx Industries, a division of Ductmate Industries
Portsmouth, VA

Craig Wray, PEng
Member ASHRAE
Consulting Engineer
Winnipeg, Manitoba, Canada

Richard Evans
Member ASHRAE
Retired
Richland, WA

Project Monitoring Subcommittee

Pat Brooks, PE (Chair)
SMACNA
Cincinnati, OH

Larry Smith, PE
Linx Industries, a division of Ductmate Industries
Portsmouth, VA

Stephen Idem, PhD
Tennessee Tech University
Cookeville, TN

Updates and errata for this publication will be posted on the ASHRAE website at www.ashrae.org/publicationupdates.

RP-1180

DUCT SYSTEMS DESIGN GUIDE

Herman Behls

ASHRAE

Peachtree Corners

ISBN 978-1-947192-40-9 (paperback)
ISBN 978-1-947192-41-6 (PDF)

180 Technology Parkway
Peachtree Corners, GA 30092
www.ashrae.org

Cover photograph iStock.com/OGGM.

Library of Congress Cataloging-in-Publication Data

Names: Behls, Herman, 1929-2017 author. | ASHRAE (Firm), editor.
Title: ASHRAE design guide for duct systems / Herman Behls.
Other titles: Design guide for duct systems
Description: Atlanta, GA : ASHRAE, [2020] | Includes bibliographical references. | Summary: "Provides engineers and other design professionals with the tools and knowledge to design duct systems so that fans are properly sized, minimize the installed cost of the ductwork, minimize system-generated noise, and minimize the fan energy consumption of commercial and industrial duct systems"-- Provided by publisher.
Identifiers: LCCN 2020056275 | ISBN 9781947192409 (paperback) | ISBN 9781947192416 (pdf)
Subjects: LCSH: Air ducts--Design and construction.
Classification: LCC TH7683.D8 B44 2020 | DDC 697.9/2--dc23
LC record available at https://lccn.loc.gov/2020056275

✧ Dedication ✧

Duct Systems Design Guide is dedicated to the lasting legacy of Herman Behls.

When ASHRAE RP-1180 could not fulfill the vision for this guide, Herman volunteered to take on this responsibility. Sadly, Herman passed away on July 12, 2017, after a brief illness, and never got to witness the book's completion. To the very end Herman was working every day to finish its writing, and he is given full credit for the contents of this book.

Herman was a former chair of ASHRAE Technical Committee (TC) 5.2 and served as the Research Subcommittee Chair for many years. Herman also served as a member of the Standards Committee and the Research Administration Committee. The contributions Herman made to the ASHRAE community are almost too numerous to count, but several honors he received deserve special mention. Herman became a member of the ASHRAE College of Fellows in 1995. This is reserved for ASHRAE members who have attained distinction in the fields of heating, refrigeration, air conditioning, ventilation, or the allied arts and sciences through invention, research, teaching, design, or original work or as an engineering executive on projects of unusual or important scope. Herman was also awarded the Service to ASHRAE Research Award in 2013 in recognition of his many years of contributions to the ASHRAE research program. At the time of this writing, Herman was one of only eleven recipients of this award since its inception.

Herman was the author of numerous work statements, acted as a PMS chair for many projects, reviewed papers, made seminar presentations, and always inspired thoughtful inquiry. In 2016 Herman was further honored with the Distinguished 50-Year Member Award, which recognizes persons who have been members of ASHRAE for 50 years and have performed outstanding service to the Society.

Among his other accomplishments, Herman was a principal author of ASHRAE Standards 120, 126, 130, and 215. He also was the primary author for the Duct Design chapter in *ASHRAE Handbook—Fundamentals* for several decades. Herman is also the father of the ASHRAE Duct Fitting Database and spent countless hours maintaining it. He was always willing to serve as a resource for anyone needing answers to their questions about the program.

Herman never stopped acquiring new knowledge from others and he was always willing to share his expertise with anyone who wanted to learn from a true teacher, friend, and mentor.

Contents

Includes spreadsheets in Microsoft® Excel® format for designing commercial and industrial duct systems by the equal friction, static regain, and constant velocity methods at www.ashrae.org/DuctSyst.

Preface

The goal of this design guide is to provide engineers and other design professionals with the tools and knowledge to design duct systems so that fans are properly sized, minimize the installed cost of the ductwork, minimize system-generated noise, and minimize fan energy. The original vision was that this guide would clearly describe procedures useful for entry-level engineers and designers who were first learning about duct design fundamentals. However, it was also desired that the guide be a valuable tool for more experienced engineers and design professionals wishing to refresh their design knowledge.

The guide was initially conceived as ASHRAE Research Project RP-1180. The Project Monitoring Subcommittee (PMS), under the auspices of ASHRAE Technical Committee (TC) 5.2, Duct Design, took control of the editorial process and accepted the task of authoring the guide after it became apparent that writing a duct design guide suitable for these disparate audiences was exceptionally challenging. Herman Behls, a former chair of TC 5.2, volunteered to complete the writing.

Introduction

OVERVIEW

Ducts in air distribution systems are designed to transport a given volume of air as efficiently and quietly as possible to minimize energy consumption and eliminate excessive noise. Duct system layout, fitting selection, leakage, acoustics, and equipment selection are important aspects of duct design.

The goals of duct design are

- to design energy-efficient HVAC systems that deliver the proper quantity of air to specific areas of a building or zones while minimizing first costs and operating costs and
- to provide a comprehensive design and a high-quality project with adequate funding and minimum construction disputes to the owner.

Various aspects of the duct design process include minimizing fan energy use, as depicted in the figure below. All projects should document the Owner's Project Requirements (OPR)—the importance of this cannot be overemphasized. The OPR should be reviewed throughout the design but especially during the commissioning process. (A comprehensive discussion of commissioning is beyond the scope of this book, but there are numerous sources of information available, such as those listed at www.ashrae.org/commissioning-essentials.)

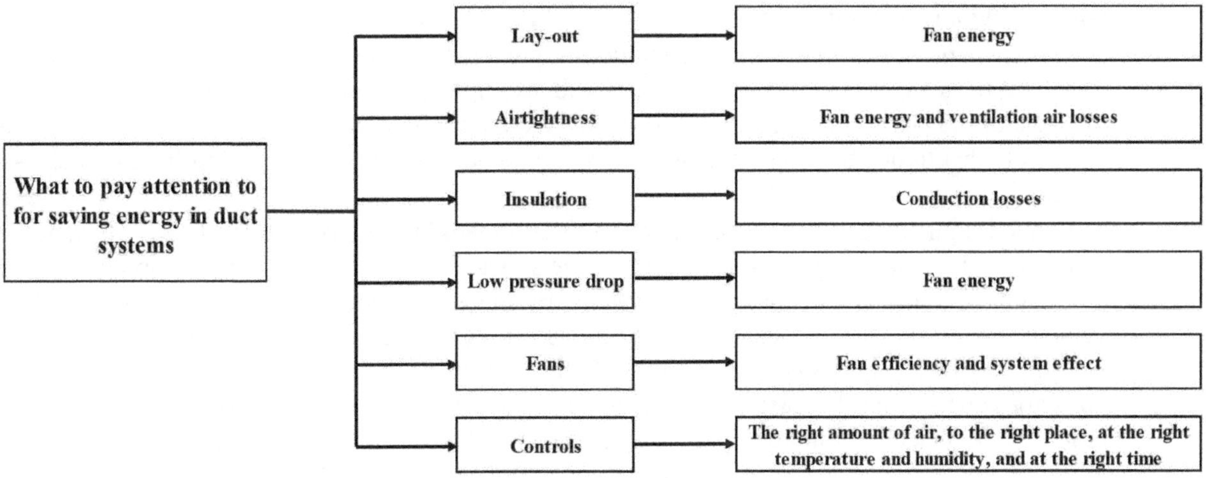

Energy-Saving Opportunities in Duct Systems

SCOPE

This design guide covers commercial and industrial duct systems. Residential duct system design is covered by Air Conditioning Contractors of America's (ACCA) *Manual D* (2016).

This guide does not cover thermal gravity effects (stack effect) or other similar applications where the duct system is vertical and density changes will impact the overall design. For stack effect examples and calculations, consult the System Analysis section of Chapter 21, "Duct Design," of *ASHRAE Handbook—Fundamentals* (2017c).

To determine system air quantities for commercial systems, consult Chapter 18, "Nonresidential Cooling and Heating Load Calculations," of *ASHRAE Handbook—Fundamentals* (2017a). To determine system air quantities for exhaust systems, use this design guide in conjunction with the American Conference of Governmental Industrial Hygienists (ACGIH®) *Industrial Ventilation: A Manual of Recommended Practice for Design*, 30th Edition (2019). Chapter 13 of this ACGIH manual includes the designs and air quantity requirements for numerous industrial applications.

To design room air distribution systems, consult the relevant ASHRAE Handbooks. For example, Chapter 20, "Space Air Diffusion," of *ASHRAE Handbook—Fundamentals* (2017b) presents the fundamental characteristics of space air diffusion, and Chapter 58, "Indoor Airflow Modeling," of *ASHRAE Handbook—HVAC Applications* (2019b) addresses air terminal units and their effects on occupant comfort along with tools for designing air distribution systems.

This design guide does not need to be read cover to cover. Readers with a fundamental understanding of duct design may only need to reference Chapters 3, 4, and 5 covering the equal friction, static regain, and constant velocity design methods. It is strongly suggested that readers also review the HVAC System Air Leakage section of Chapter 21, "Duct Design," of *ASHRAE Handbook—Fundamentals* (2017c), as it provides insight on system sealing requirements, scope of testing, acceptance criteria, recommended specifications, and the responsibilities of engineers and contractors.

Chapter 2 of this design guide includes a discussion on uncertainty and error considerations that readers should familiarize themselves with to appreciate the pressure drop calculation process.

Chapter 8 of this book offers a comprehensive overview of duct system acoustics; it is recommended that for additional guidance on acoustics readers see *Noise and Vibration Control* (Schaffer 2005) as well as Chapter 49, "Noise and Vibration Control," of *ASHRAE Handbook—HVAC Applications* (2019a). Schaeffer addresses general noise-control design guidelines during the various design phases and includes information on the architectural and structural aspects of HVAC system design. Sample specifications for acoustical materials and the acoustical performance of HVAC equipment are also provided. Schaffer also gives suggestions for troubleshooting HVAC noise and vibration complaints.

Each chapter of this design guide includes a nomenclature section at the end that defines the variables used in the chapter's equations, and example problems and solutions are included throughout the book to help readers understand the concepts of duct design with practical examples. Note that throughout this book all values and nomenclature are provided in Inch-Pound (I-P) units only.

This design guide is also accompanied by free spreadsheets in Microsoft® Excel® format that can be found at www.ashrae.org/DuctSyst. These spreadsheets can be used for designing commercial, industrial, and local exhaust duct systems by the equal friction, static regain, and constant velocity methods. The spreadsheets are best used in conjunction with the web-based ASHRAE Duct Fitting Database (DFDB), available via subscription at www.ashrae.org/technical-resources/bookstore/duct-fitting-database (2016). This online database includes duct friction losses and fitting loss coefficients for more than 200 round, rectangular, and flat oval duct fittings and is useful to design engineers in the calculation of friction factors and loss coefficients for a wide variety of ducts and fittings.

REFERENCES

ACCA. 2016. ANSI/ACCA Manual D-2016, *Manual D®—Residential duct systems*. Arlington, VA: Air Conditioning Contractors of America.

ACGIH. 2019. *Industrial ventilation: A manual of recommended practice for design*, 30th ed. Cincinnati, OH: American Conference of Governmental Industrial Hygienists.

ASHRAE. 2016. ASHRAE Duct Fitting Database, ver. 6.00.05 (online). Peachtree Corners, GA: ASHRAE. www.ashrae.org/technical-resources/bookstore/duct-fitting-database.

ASHRAE. 2017a. Chapter 18, Nonresidential cooling and heating load calculations. In *ASHRAE handbook—Fundamentals*. Peachtree Corners, GA: ASHRAE.

ASHRAE. 2017b. Chapter 20, Space air diffusion. In *ASHRAE handbook—Fundamentals*. Peachtree Corners, GA: ASHRAE.

ASHRAE. 2017c. Chapter 21, Duct design. In *ASHRAE handbook—Fundamentals*. Peachtree Corners, GA: ASHRAE.

ASHRAE. 2019a. Chapter 49, Noise and vibration control. In *ASHRAE handbook—HVAC Applications*. Peachtree Corners, GA: ASHRAE.

ASHRAE. 2019b. Chapter 58, Indoor airflow modeling. In *ASHRAE handbook—HVAC Applications*. Peachtree Corners, GA: ASHRAE.

Schaffer, M.E. 2005. *A practical guide to noise and vibration control for HVAC systems*, 2nd ed. (I-P). Peachtree Corners, GA: ASHRAE.

1

Duct Design Fundamentals

OVERVIEW

This chapter covers the fundamentals of duct design: pressure losses for round, rectangular and flat oval ducts and fittings and the two fundamental physical laws of conservation of mass and conservation of energy, which govern the flow of air in ducts. These laws permit the derivation of the continuity and pressure equations, which are the basis for duct system design. The chapter also discusses the ASHRAE Duct Calculator (a spreadsheet tool available with this book that can be used to evaluate pressure loss applications per duct section), friction charts, and the online tool ASHRAE Duct Fitting Database (DFDB).

PRESSURE LOSS

The pressure loss for each section of a duct system can be determined using the Darcy-Weisbach equation, which says the total pressure loss is the sum of the friction loss of the ductwork and the dynamic loss of the components (fittings, obstructions, etc.):

$$\Delta p_t = \Delta p_{friction} + \Delta p_{dynamic}$$

The Darcy-Weisbach equation can be rewritten showing the friction and dynamic loss in a particular duct section as Equation 1-1:

$$\Delta p_t = \left(\frac{12fL}{D_h} p_v\right) + Cp_v \qquad (1\text{-}1)$$

where C is a fitting loss coefficient from the DFDB and p_v is the velocity pressure, which can be calculated by hand or calculated and displayed in the DFDB.

This equation consists of two terms on the right side: the Darcy equation (the first term), which calculates the resistance to airflow caused by friction, and the Weisbach equation (the second term), which calculates dynamic losses in the section. The Darcy equation can be written as

$$\Delta p_{friction} = \left(\frac{12fL}{D_h} p_v\right) \qquad (1\text{-}2)$$

This equation can be calculated by hand. However, it is easier to get this value from the ASHRAE Duct Fitting Database (DFDB; ASHRAE 2016) or a friction chart like that available in Chapter 21, "Duct Design," of *ASHRAE Handbook—Fundamentals* (2017a). The friction rate (pressure loss per 100 ft of duct) determined from the chart must be multiplied by the duct length in feet and then divided by 100 to determine the friction loss.

This chapter discusses the fundamentals of airflow and the supporting equations needed to solve Equation 1-1. When manually calculating the two components of Equation 1-1, it is best to use a spreadsheet with the methodology presented for breaking a system into duct sections such that fitting loss coefficients in each section are additive. Sectioning a duct system is discussed in Chapter 2.

CONSERVATION OF MASS

To establish proper flow rate of air within ducts, the conservation of mass equation must be used. The law of conservation of mass for a steady flow states that the mass flow rate flowing into a duct must equal the mass flow rate out of that section of duct if no mass is added or lost (e.g., leakage). As an equation, this is written as follows:

$$\dot{m} = \rho A_d V = \text{constant}$$

Continuity Equation

When air density is constant in a duct system, the volumetric flow rate at any duct section is

$$Q = A_d V$$

where duct area is calculated in square feet as follows:

Round: $$A_d = \frac{\pi}{4}(D/12)^2$$

Rectangular: $$A_d = (W/12)(H/12)$$

Flat oval: $$A_d = (\pi(a/12)^2/4) + (a/12)((A/12) - (a/12))$$

Variations of the continuity equation are as follows:

- Knowing volumetric flow Q and duct cross-sectional area, the duct velocity V can be calculated using Equation 1-3:

$$V = \frac{Q}{A_d} \tag{1-3}$$

- Knowing volumetric flow Q and duct velocity V, the duct cross-sectional area can be calculated using the following equation:

$$A_d = \frac{Q}{V}$$

Examples 1-1 and 1-2 work through variations in the continuity equation. Example 1-3 shows how to determine the duct diameter with known values for volume flow rate and velocity.

Example 1-1.
If the average velocity in a 22 in. diameter round duct is measured and found to be 1850 fpm, what is the volume flow rate?

Solution.

$$A_d = \frac{\pi}{4}(D/12)^2 = \frac{\pi}{4}(22/12)^2 = 2.64 \text{ ft}^2$$

$$Q = A_d V = (2.64)(1850) = 4884 \text{ cfm}$$

Example 1-2.
If the volume flow rate in a 22 in. diameter round duct is 5000 cfm, what is the average velocity of air in the duct?

Solution.

$$A_d = \frac{\pi}{4}(D/12)^2 = \frac{\pi}{4}(22/12)^2 = 2.64 \text{ ft}^2$$

$$V = \frac{Q}{A_d} = \frac{5000}{2.64} = 1894 \text{ fpm}$$

Example 1-3.
If the design volume flow rate and velocity are 13,000 cfm and 4000 fpm minimum, respectively, what is the duct diameter?

Solution.

$$A_d = \frac{Q}{V} = \frac{13{,}000}{4000} = 3.25 \text{ ft}^2$$

$$D = 12\sqrt{\frac{4A_d}{\pi}} = 12\sqrt{\frac{(4)(3.25)}{\pi}} = 24.4 \text{ in.}^2$$

Diverging Flow

According to the law of conservation of mass, the volume flow rate before flow divergence is equal to the sum of the flows after divergence. Equation 1-4 and Figure 1-1 illustrate this point.

$$Q_c = Q_h + Q_s \tag{1-4}$$

Converging Flow

According to the law of conservation of mass, the volume flow rate after flow convergence is equal to the sum of the flows before convergence. Equation 1-4 and Figure 1-2 illustrate this point.

CONSERVATION OF ENERGY

Pressure losses in a duct system due to friction and dynamic effects must be compensated by energy added to the flow by means of a fan. To understand how pressure losses are determined, the principle of conservation of energy must be employed.

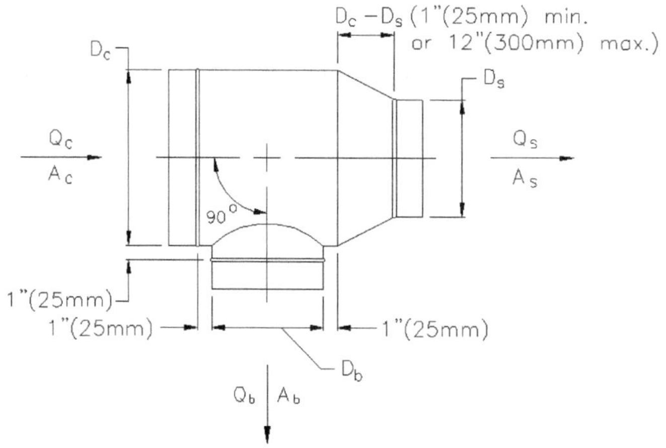

Figure 1-1 Diverging Flow Tee
(Reprinted from ASHRAE 2016, fitting SD5–9)

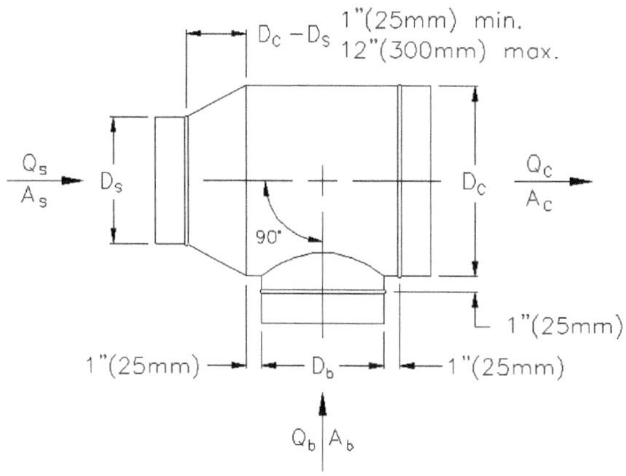

Figure 1-2 Converging Flow Tee
(Reprinted from ASHRAE 2016, fitting ED5–3)

The total energy per unit volume of airflow at any point in a duct system is equal to the sum of the static energy, kinetic energy, and potential energy. When applied to airflow in ducts, the static energy (flow work) is represented by the static pressure of the air, and the kinetic energy is represented by the velocity pressure. Potential energy is due to elevation above a reference datum and is often negligible in HVAC duct systems. Consequently, the total pressure (total energy) of air flowing in a duct system is generally equal to the sum of the static pressure and the velocity pressure. As an equation, this is written as follows:

$$p_t = p_s + p_v$$

From the conservation of energy equation written for steady incompressible flow and a negligible change in elevation, the change in static pressure between any two points of a system is equal to the sum of the change in static pressure and the change in velocity pressure. This relationship is represented by the following equation:

$$\Delta p_t = \Delta p_s + \Delta p_v$$

Pressure (or pressure loss) is important to all duct designs and sizing methods. Many times, systems are sized to operate at a certain pressure or not in excess of a certain pressure. Higher pressure at the same volume flow rate means that more energy is required from the fan. This increase in energy requirements increases the system operating cost.

Static Pressure

Static pressure is a measure of the static energy of air flowing in a duct system. It is termed *static* in that it can exist without movement of the airstream. The air that fills a balloon is a good example of static pressure: the pressure is exerted equally in all directions, and the magnitude of the pressure is reflected by the size of the balloon.

The atmospheric pressure of air is a static pressure. At sea level, this pressure is equal to 14.696 psia. For air to flow in a duct system a pressure differential must exist; energy must be imparted to the system by a fan or other device to raise the pressure above or below atmospheric pressure.

Air always flows from an area of higher pressure to an area of lower pressure. The static pressure is above atmospheric pressure at a fan outlet, so air flows from the fan through any connecting ductwork until it reaches atmospheric pressure at the discharge. The static pressure is below atmospheric at a fan inlet, so air flows from the higher atmospheric pressure through an intake and any connecting ductwork until it reaches the area of lowest static pressure at the fan inlet. The first type of system is referred to as a *positive pressure* or *supply air* system, and the second type is referred to as a *negative pressure system*, which can be an *exhaust* or *return air* system.

Sign Convention

When static pressure is expressed as a positive number, it means the pressure is greater than the atmospheric pressure. Negative static pressure indicates a pressure less than atmospheric pressure.

By convention, positive changes in static pressure represent losses and negative changes represent regains or increases. For example, if the static pressure change as air flows from point A to point B in a system is a positive number, then there is a static pressure loss between points A and B and the static pressure at point A is greater than the static pressure at point B. Conversely, if the static pressure change as air flows between these points is negative, the static pressure at point B is greater than the static pressure at point A and there is a regain. Figure 1-3 illustrates static pressure losses and regains. Static pressure regains occur between sections 2 and 3 and sections 5 and 6.

Velocity Pressure

Velocity pressure is a measure of the kinetic energy of the air flowing in a duct system. It is directly proportional to the square of the air velocity:

$$p_v = \rho\left(\frac{V}{1097}\right)^2 \tag{1-5a}$$

In Equation 1-5a, the quantity 1097 is a units conversion factor. For standard air density, substituting $\rho = 0.075\ \text{lb}_m/\text{ft}^3$ into Equation 1-5a yields the following:

$$p_v = \left(\frac{V}{4005}\right)^2 \tag{1-5b}$$

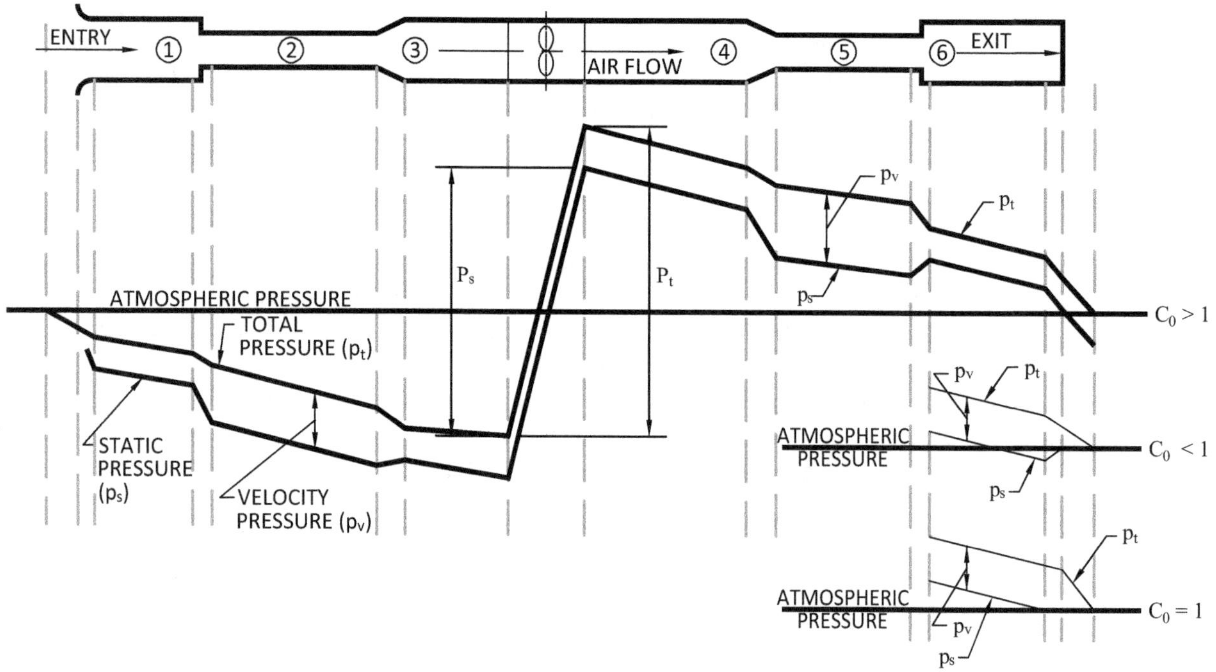

Figure 1-3 Pressure Changes During Flow in Ducts
(Adapted from ASHRAE 2017, Figure 7)

Velocity pressure is always a positive number. From Equation 1-3 ($V = Q/A_d$), it can be seen that velocity must increase if the duct diameter (area) is reduced without a reduction in air volume. Similarly, the velocity must decrease if the air volume is reduced without a corresponding reduction in duct diameter. Thus, velocity and velocity pressure in a duct system are constantly changing (see Figure 1-3). If the velocity pressure change as air flows from one point to another point in a system decreases, then there can be a corresponding increase in static pressure. This relationship is called *static pressure regain* and is illustrated in Figure 1-3.

Total Pressure

The energy of air flowing in a duct system is represented by total pressure. As there is no way to create or increase energy except to add heat or work, the total pressure cannot be increased once air leaves the fan. Therefore, the total pressure at the fan outlet is at the maximum value, and the pressure decreases continually as the air moves toward the outlets. Losses in total pressure represent the conversion of kinetic energy and static energy to internal energy in the form of heat and are classified either as dynamic losses or as friction losses. Dynamic loses result from duct system changes in shape, size, direction, or volume flow rate or from turbulence. Friction result from moving air flowing in contact with a fixed boundary (McGill 2003).

Pitot-Static Tube

A pitot-static tube (Figure 1-4) is a device used to measure the local air velocity. It consists of two coaxial tubes pointing directly into the flow. An open port located at the tip of the inner tube of the probe senses the total air pressure. Several pressure ports oriented perpendicular to the flow are equally spaced about the circumference of the outer tube; these ports sense the static pressure. The difference between the two pressures, which is used to evaluate the velocity pressure, is typically measured using a manometer or a pressure transducer.

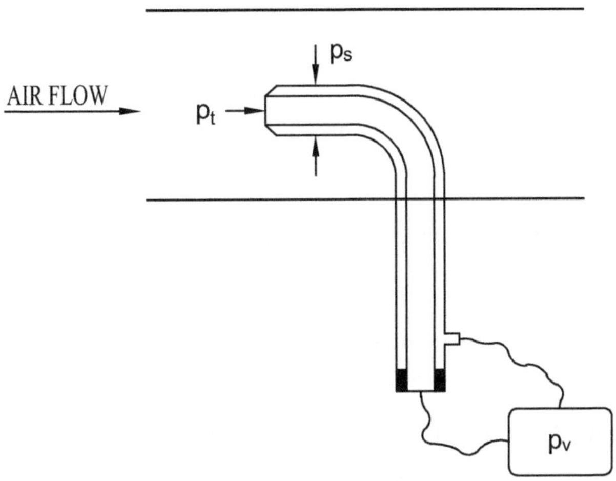

Figure 1-4 Pitot-Static Tube Showing Static and Velocity Pressures in a Duct

DUCT DESIGN SPREADSHEET BASED ON MANUAL CALCULATIONS

The spreadsheet tool ASHRAE Duct Calculator, available with this book at www.ashrae.org/DuctSyst, can be used to evaluate pressure loss applications per duct section, with the DFDB (ASHRAE 2016) used to calculate required fitting loss coefficients. The DFDB can also be used to calculate friction loss in ducts. Though friction charts (which are approximate) and fundamental equations will provide friction loss data, the DFDB offers the most complete compilation of fitting loss coefficients available, and its use is recommended.

Either method permits the solution of the Darcy-Weisbach equation (Equation 1-1). As previously stated, there are two components of this equation. The first term on the right side is the Darcy equation, which calculates the resistance to airflow caused by friction. The second term on the right side is the Weisbach equation, which calculates dynamic losses in the section. The equation calculates the pressure loss sum attributed to each section of duct in the design leg (which is the particular pathway in the layout that exhibits the maximum pressure loss, i.e., the largest flow resistance).

Solving the Darcy Equation

To calculate the resistance of straight duct, including its friction factor *f*, the following equations are needed:

$$D_h = \frac{12(4A_d)}{P}$$

The duct area A_d and perimeter P are calculated as follows:

Round: $$A_d = \frac{\pi}{4}(D/12)^2 \quad ; \quad P = \frac{\pi D}{12}$$

Rectangular: $$A_d = \frac{WH}{144} \quad ; \quad P = \frac{2(W+H)}{12}$$

Flat oval: $$A_d = \frac{(\pi a^2/4) + a(A-a)}{144} \quad ; \quad P = \frac{\pi a + 2(A-a)}{12}$$

The friction factor is calculated using the Colebrook equation (Equation 1-6):

$$\frac{1}{\sqrt{f}} = -2\log\left(\frac{12\varepsilon}{3.7D_h} + \frac{2.51}{\text{Re}\sqrt{f}}\right) \tag{1-6}$$

where the dimensionless Reynolds number is defined as

$$\text{Re} = \frac{\rho V D_h}{12\mu}$$

where the average air velocity in a duct section is given by Equation 1-3 ($V = Q/A_d$) and where air density (ρ) and dynamic viscosity (μ) at standard conditions are obtained from Chapter 3, "Fluid Flow," of *ASHRAE Handbook—Fundamentals* (2017b) and actual conditions are obtained using a calculator with psychometric routines or from the DFDB (ASHRAE 2016) by selecting the "show air properties" radio button in the output section of the DFDB.

The Colebrook equation (Equation 1-6) cannot be solved explicitly for f—the solution must be found iteratively. The variable f is the friction factor and the variable ε is the absolute duct roughness, which must be determined empirically. Typical values of ε are given in Table 1-2 in the ASHRAE Duct Fitting Database (DFDB) section of this chapter. The spreadsheet tool ASHRAE Duct Calculator, available with this book at www.ashrae.org/DuctSyst, will solve the Colebrook equation for f and calculate air properties, friction losses of round and flexible ducts, and equivalent round sizes of rectangular and flat oval ducts. The DFDB (ASHRAE 2016) will also calculate air properties and friction losses using these equations. Use of the DFDB is discussed in the following ASHRAE Duct Fitting Database (DFDB) section.

The velocity pressure is evaluated using Equation 1-5b, and the friction pressure loss is determined using the Colebrook equation (Equation 1-6) in conjunction with the Darcy equation (Equation 1-2).

Examples 1-4 and 1-5 work through friction loss determination, and Example 1-6 shows pressure loss determination.

Determining Equivalent Duct Sizes

The following equations are used to determine the rectangular and flat oval duct size equivalents to round duct for equal airflow, length, and resistance, or vice versa, for a known equivalent circular diameter D_e. Because there are an infinite number of duct dimension combinations for each D_e, it is more convenient to use the tables in the Equivalent Round Tables spreadsheet available with this book at www.ashrae.org/DuctSyst than to attempt to calculate the equivalents manually.

Rectangular:
$$D_e = \frac{1.30AR^{0.625}}{p^{0.250}} = \frac{1.30(WH)^{0.625}}{(W+H)^{0.250}} \tag{1-7}$$

Flat oval:
$$D_e = \frac{1.55AR^{0.625}}{p^{0.250}} = \frac{1.55\left[\frac{\pi}{4}a + a(A-a)\right]^{0.625}}{[\pi a + 2(A-a)]^{0.250}} \tag{1-8}$$

Example 1-4.
Determine the friction loss of 20 ft of 22 in. diameter spiral round duct with 10 ft joints at an airflow rate of 5000 cfm for Denver, Colorado, with an elevation of 5400 ft above sea level.

Solution.

ρ = 0.061 lb_m/ft^3 (taken from the DFDB)
μ = 0.000734 $lb_m/(ft\cdot min)$ (taken from the DFDB)
ε = 0.0003 ft (see Table 1-2)
D_h = 22 in. (Note: For a round duct, $D_h = D$)

$$A_d = \frac{\pi}{4}(D/12)^2 = \frac{\pi}{4}(22/12)^2 = 2.64 \text{ ft}^2$$

$$V = \frac{Q}{2.64} = \frac{5000}{2.64} = 1894 \text{ fpm}$$

$$\text{Re} = \frac{\rho V D_h}{12\mu} = \frac{(0.061)(1894)(22)}{(12)(0.000734)} = 2.89 \times 10^5$$

$$\text{Friction factor} = \frac{1}{\sqrt{f}} = -2\log\left(\frac{12\varepsilon}{3.7D_h} + \frac{2.51}{\text{Re}\sqrt{f}}\right) = -2\log\left(\frac{(12)(0.0003)}{(3.7)(22)} + \frac{2.51}{(2.89 \times 10^5)\sqrt{f}}\right)$$

where $f = 0.0160$ (an iterative solution; see ASHRAE Duct Calculator available with this book online).

$$p_v = \rho\left(\frac{V}{1097}\right)^2 = 0.061\left(\frac{1894}{1097}\right)^2 = 0.18 \text{ in. of water}$$

$$\Delta p_f = \left(\frac{12fL}{D_h}\right)p_v = \left(\frac{(12)(0.0160)(20)}{22}\right)(0.18) = 0.03 \text{ in. of water}$$

FRICTION CHART

A friction chart (see Figure 1-5) is a graph that shows the friction loss per defined length (100 ft) for a given duct diameter at a specific air quantity and air density. Friction charts are based on the Colebrook equation (Equation 1-6), the duct roughness factors listed in Table 1-2, and the Darcy equation (Equation 1-2). The duct roughness factors account for the inherent surface roughness of the duct wall as well as the additional resistance due to transverse joints.

Figure 1-5 is a friction chart for standard air and an absolute roughness $\varepsilon = 0.0003$ ft. This standard friction chart can be used for the conditions listed below. Within these constraints the result of the friction chart is within ±5% of actual losses.

- Medium smooth ducts per Table 1-2
- Temperature variations not exceeding 70°F±30°F
- Duct pressures from –20 in. of water to +20 in. of water relative to ambient pressure
- Elevations up to 1500 ft

Example 1-5.
Determine the friction loss of 30 ft of 26 × 16 in. galvanized rectangular duct with 4 ft joints at an airflow rate of 5000 cfm for Denver, Colorado. Denver's elevation is 5400 ft.

Solution.

ρ = 0.061 lb_m/ft^3 (taken from the DFDB)
μ = 0.000734 $lb_m/(ft \cdot min)$ (taken from the DFDB)
ε = 0.0003 ft (see Table 1-2)

$$A_d = \frac{WH}{144} = \frac{(26)(16)}{144} = 2.89 \text{ ft}^2$$

$$P = \frac{2(H+W)}{12} = \frac{2(16+26)}{12} = 7.0 \text{ ft}$$

$$D_h = \frac{12(4A_d)}{P} = \frac{(12)(4)(2.89)}{7.0} = 19.8 \text{ in.}$$

$$V = \frac{Q}{A_d} = \frac{5000}{2.89} = 1730 \text{ fpm}$$

$$\text{Re} = \frac{\rho V D_h}{12\mu} = \frac{(0.061)(1730)(19.8)}{(12)(0.000734)} = 2.37 \times 10^5$$

$$\text{Friction factor} = \frac{1}{\sqrt{f}} = -2\log\left(\frac{12\varepsilon}{3.7D_h} + \frac{2.51}{\text{Re}\sqrt{f}}\right) = -2\log\left(\frac{(12)(0.0003)}{(3.7)(19.8)} + \frac{2.51}{(2.37 \times 10^5)\sqrt{f}}\right)$$

where $f = 0.0166$ (an iterative solution; see ASHRAE Duct Calculator available with this book online).

$$p_v = \rho\left(\frac{V}{1097}\right)^2 = 0.061\left(\frac{1730}{1097}\right)^2 = 0.15 \text{ in. of water}$$

$$\Delta p_f = \left(\frac{12fL}{D_h}\right)p_v = \left(\frac{(12)(0.0166)(30)}{19.8}\right)(0.15) = 0.05 \text{ in. of water}$$

For conditions outside these constraints use the DFDB (ASHRAE 2016). (Though duct calculators may be used to achieve approximate results, they are subject to calculation errors if these constraints are violated.)

Standard air is defined as air at a density of 0.075 lb_m/ft^3, which is approximately 70°F and 0% rh, or 68°F and 50% rh, at a sea level pressure of 14.696 psi. Normally, air pressure variations within a duct system caused by pressure losses are ignored and the density is determined at the surrounding atmospheric pressure.

ASHRAE DUCT FITTING DATABASE (DFDB)

The ASHRAE Duct Fitting Database (DFDB) is a simple-to-use, windows-based database that enables users to select from over 200 fittings and enter information such as airflow and size then calculates velocity, velocity pressure, loss coefficient, and pressure loss (ASHRAE 2016). For junctions, the loss coefficients are given for the main straight-through path and the branch path.

Example 1-6.

Determine the pressure loss in the three-component system shown in the figure below, which is comprised of a transition between two ducts having different cross-sections. The system is located in Denver, Colorado, and the duct material is galvanized steel. Round duct is spiral, and the rectangular duct has 4 ft joints. The 26 × 16 in. duct is 30 ft in length; the 22 in. duct is 20 ft long. Airflow is 5000 cfm.

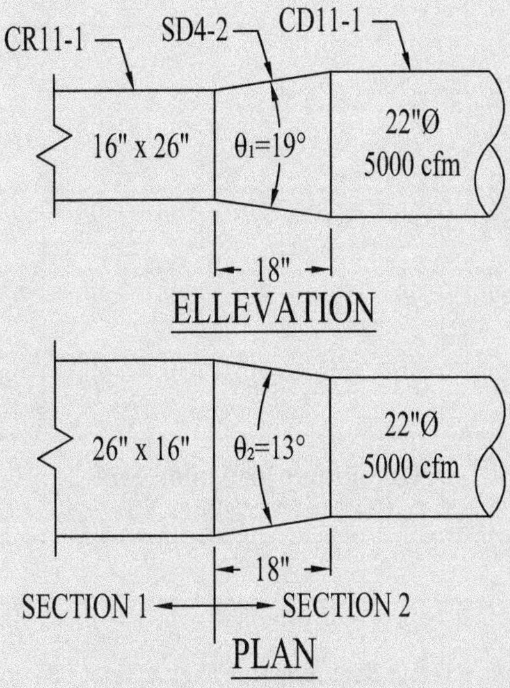

Flow Through a Transition

Solution.

Table 1-2 shows the material roughness category is medium smooth ($\varepsilon = 0.0003$ ft). For Denver, at elevation 5400 ft, the air density is 0.061 lb_m/ft^3. Dynamic viscosity μ is 7.34×10^{-4} lb_m/ft·min.

The figure below summarizes the calculations. Because the system is a supply air system, the transition fitting is assigned to downstream section 2.

Equal Friction Duct Design Example - 3500 fpm

Air Temperature, °F	70	Relative Humidity, %	0
Elevation, ft	5400	Air Density, lbm/ft³	0.061
Barometric Pressures, psia	12.046	Viscosity (μ), lbm/(ft-min)	0.0007335

Upstream Section	Section	Fitting	ASHRAE Fitting Code	Air Quantity, cfm	Duct Size, in.	Velocity, fpm	Duct Length, ft	Velocity Pressure, p_v, n. of water	Loss Coefficient C	Total Pressure Loss, in. of water
		Source		Source					Source	
		Drawings	DFDB	Drawing	DFDB	DFDB	Drawing	DFDB	DFDB	Σ
Fan	1	Duct - Rectangular	CR11-1	5000	26 x 16	1731	30			0.05
									0.00	0.00
Section Total										0.05
1	2	Duct - Rectangular	CD11-1	5000	22	1894	20			0.03
		Transition (see Figure 2-7)	SD4-2						0.01	
								0.18	0.01	0.00
Section Total										0.03

Summarized Calculations

For the 16 in. spiral duct, calculations refer to Example 1-4. For the 22 × 10 in. rectangular duct, calculations refer to Example 1-5. The DFDB output for fitting SD4-2 is shown in the figure below.

SD4-2 Transition, Rectangular to Round, Supply Air Systems
(Idelchik 1986, Diagram 5-27)

INPUT		
Height (H1)	in.	16
Width (W1)	in.	26
Diameter (Do)	in.	22
Length (L)	in.	18.0
Flow Rate (Q)	cfm	5,000
Density (RHO)	lbm/ft^3	0.061

Calculate

OUTPUT		
Angle (Theta1)	deg.	19
Angle (Theta2)	deg.	13
Velocity (Vo)	fpm	1,894
Velocity (V1)	fpm	1,731
Vel Pres at Vo (Pvo)	in. wg	0.18
Vel Pres at V1 (Pv1)	in. wg	0.15
Loss Coefficient (Co)		0.01
Loss Coefficient (C1)		0.01
Pressure Loss (Po)	in. wg	0.00

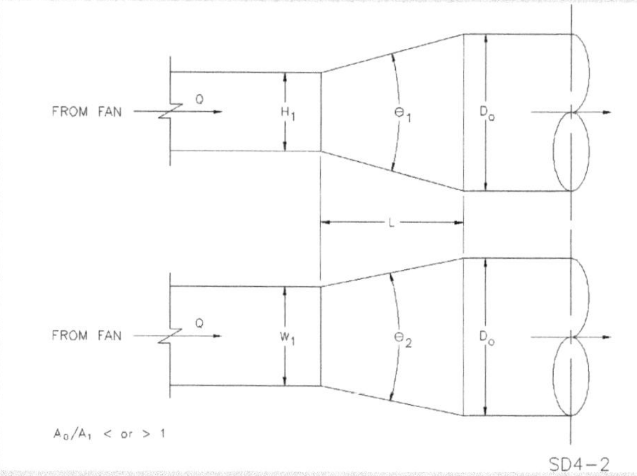

DFDB Output for Fitting SD4-2
(Reprinted from ASHRAE 2016)

DFDB Calculations

Dynamic losses happen along duct lengths. They cannot be separated from friction losses. To make calculation simpler, dynamic losses exclude friction and are assumed to be local, or concentrated in a single section. Friction losses are to be considered for longer fittings only. Friction losses of fittings are typically accounted for by measuring duct lengths between the centerline of one fitting and the centerline of the next fitting. For close-coupled fittings that are less than six hydraulic diameters apart, the flow pattern used to determine loss coefficients is different from the flow pattern entering all subsequent fittings. Close-coupled fittings should generally be avoided, as data for accurately determining pressure losses for them are unavailable (ASHRAE 2009).

Loss coefficients in the DFDB are organized by fitting type, function, and geometry. The local loss coefficient is represented by the ratio of the total pressure loss of the fitting to the local velocity pressure at the fitting:

$$C = \frac{\Delta p_{t,fitting}}{p_v}$$

DFDB Nomenclature

The DFDB (ASHRAE 2016) uses a unique nomenclature to identify each fitting/duct-mounted component. This nomenclature was developed for easy identification of fittings and is shown in Table 1-1. The fitting functions, geometries, and categories are self-explanatory with the following comments (ASHRAE 2017a):

- Entries and converging junctions are only for exhaust/return portion of systems.
- Exits and diverging junctions are only for supply systems.

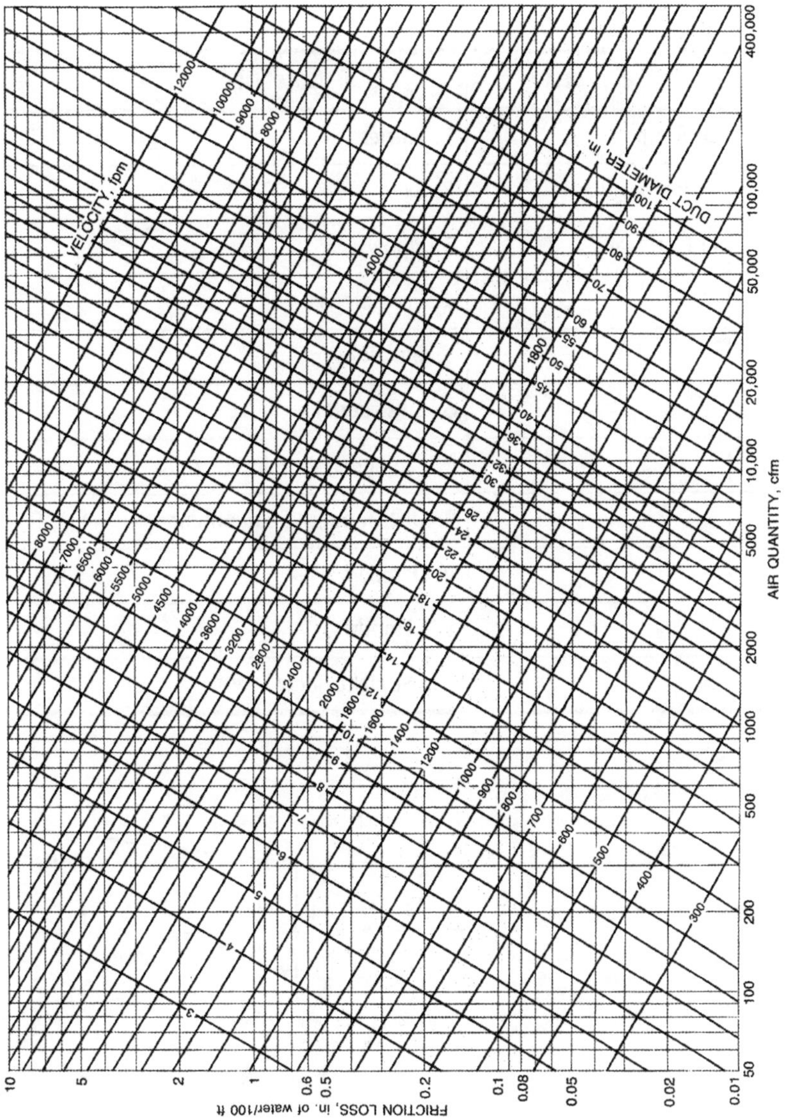

Figure 1-5 Friction Chart for Medium Smooth Ducts (ε = 0.0003 ft)
(Reprinted from ASHRAE 2017a, Figure 10)

- Equal-area elbows, obstructions, and duct-mounted equipment are common to both supply and exhaust/return systems.
- Transitions and unequal-area elbows are both supply or exhaust/return fittings, each referenced to the appropriate cross-section and direction of airflow.

Following are some examples of the nomenclature:

- Fitting ED5-1 (Figure 1-6) is an exhaust fitting (E) with a round shape (diameter, D). The numeral 5 indicates that the fitting is a junction, and the 1 indicates it is sequentially first in the database.
- Fittings SR3-1 and ER3-1 (Figure 1-7) are supply (S) and exhaust (E) fittings, respectively, with a rectangular (R) shape. The numeral 3 indicates that the fittings are elbows. These elbows have different cross-sectional areas at their entrances/exits, as shown in Figure 1-7. Therefore, their loss coefficients are unique, each associated with plane *o*.

Table 1-1 DFDB Nomenclature

Fitting Function	Geometry	Category	Sequential Number
S: Supply	D: Round	1: Entries	1, 2, 3 ... n
E: Exhaust/return	R: Rectangular	2: Exits	
C: Common	F: Flat oval	3: Elbows	
		4: Transitions	
		5: Junctions	
		6: Obstructions	
		7: Fan and system interactions	
		8: Duct-mounted equipment	
		9: Dampers	
		10: Hoods	
		11: Straight ducts	

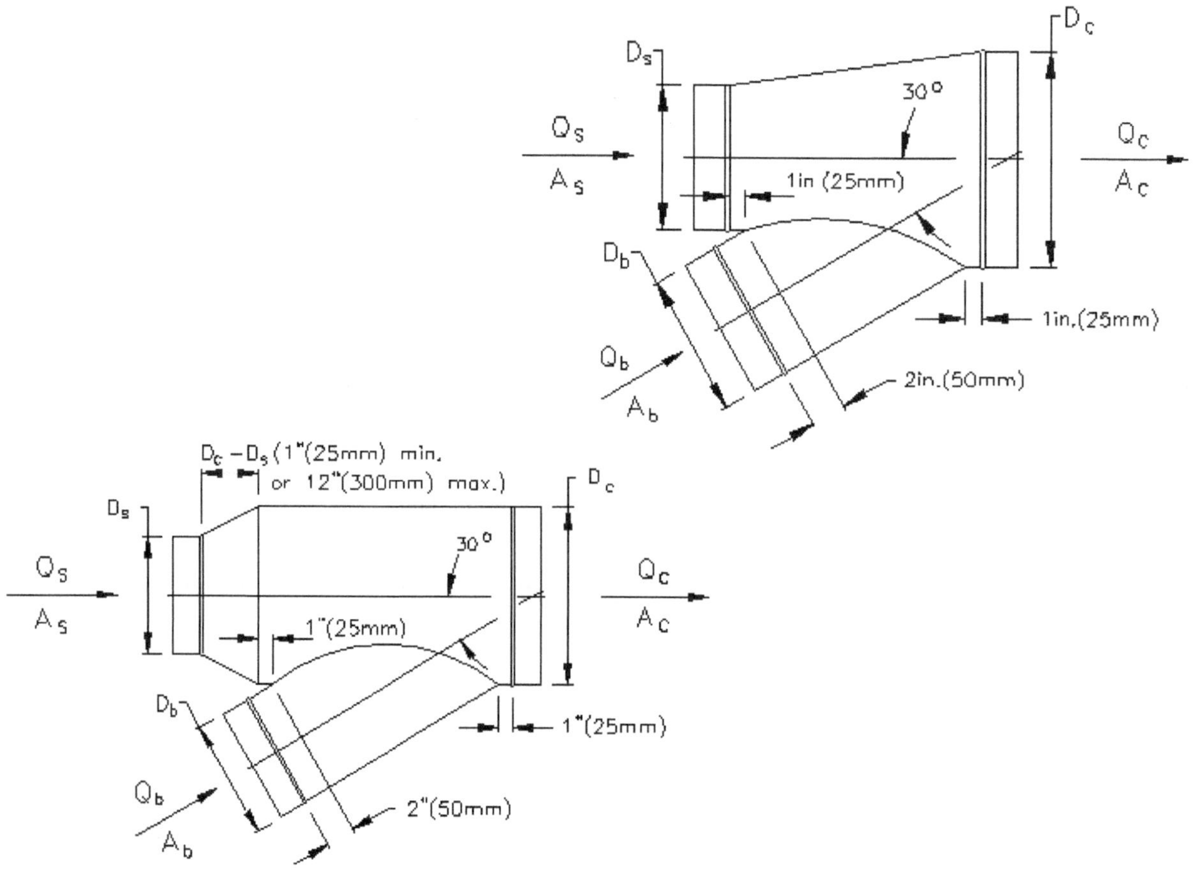

Figure 1-6 30° Converging Wye
(Reprinted from ASHRAE 2016, fitting ED5-1)

- Fitting CR3-6 (Figure 1-8) is a common (C) fitting with a rectangular (R) shape. The numeral 3 indicates that the fitting is an elbow, and the 6 indicates it is the sixth in the series in the database. This elbow is considered common as it has equal entrance and exit areas.

Loss Coefficients

Junctions, Converging and Diverging

The loss coefficients for junctions are given for the straight-through (main) path (C_s) and the branch path (C_b). The straight-through loss is reference to the downstream section s for diverging junctions and the upstream section s for converging junctions. The branch loss is referenced to the branch b for both converging and diverging junctions. The straight-through path and branch path pressure losses are calculated by the following equations, where $p_{v,s}$ and $p_{v,b}$ are the velocity pressures calculated at the straight-through (s) and branch (b) cross sections:

$$\Delta p_{t,s} = C_s \cdot p_{v,s}$$

$$\Delta p_{t,b} = C_b \cdot p_{v,b}$$

Figure 1-9 shows examples of diverging and converging tee fittings with the proper nomenclature.

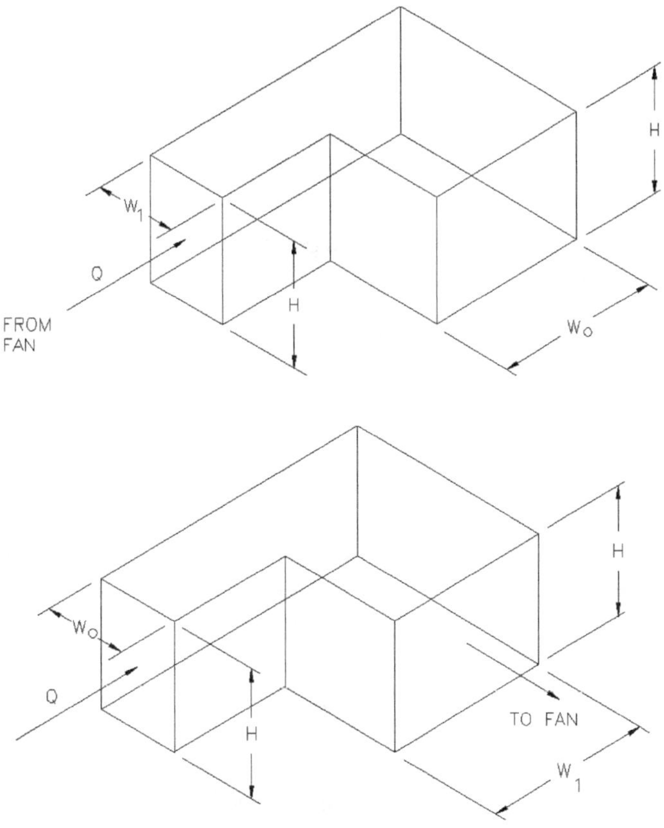

Figure 1-7 Unequal-Area Fitting Nomenclature
(Reprinted from ASHRAE 2016, fittings SR3-1 and ER3-1)

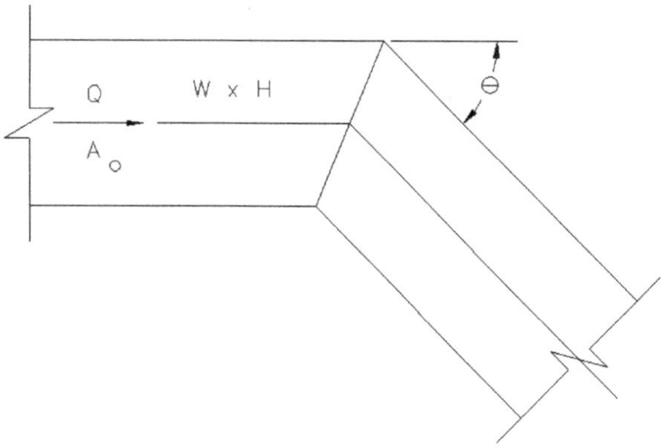

Figure 1-8 Mitered Elbow
(Reprinted from ASHRAE 2016, fitting CR3-6)

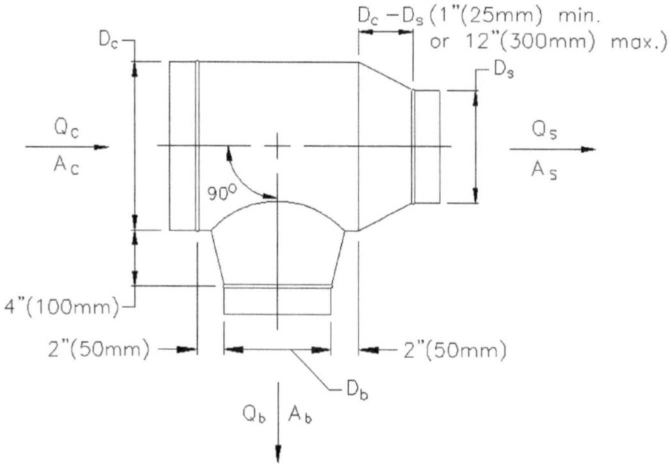

(a) Diverging-Flow Tee

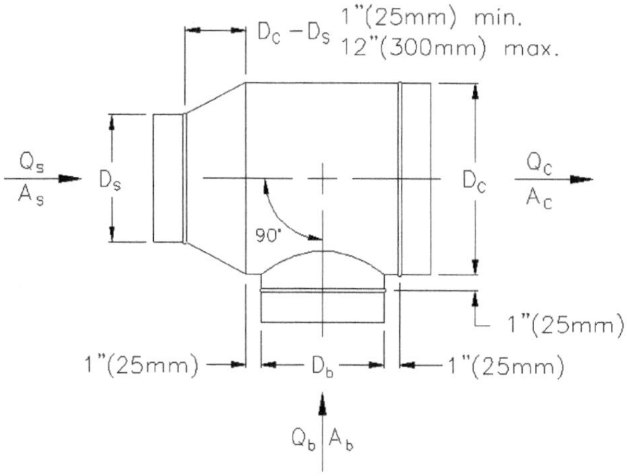

(b) Converging-Flow Tee

Figure 1-9 Examples of Nomenclature
(Reprinted from ASHRAE 2016, fittings SD5-10 and ED5-3)

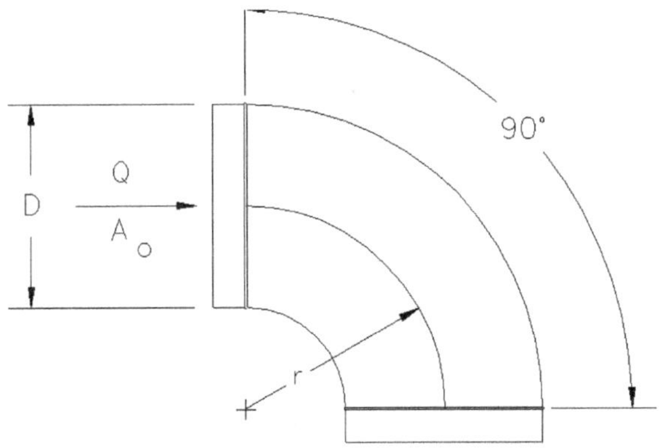

Figure 1-10 Example of an Equal Inlet/Outlet-Area Fitting
(Reprinted from ASHRAE 2016, fitting CD3-1)

Equal Inlet and Outlet Cross Sections

The pressure loss of equal inlet/outlet-area fittings is referenced to cross section *o*. The fitting total pressure loss is calculated by the following equation:

$$\Delta p_t = C_o \cdot p_{v,o}$$

Figure 1-10 shows an example of a fitting with equal inlet and outlet areas.

Unequal Inlet and Outlet Cross Sections

Figure 1-11 shows fittings with unequal inlets and outlets. Note the reference section (where velocity pressure is calculated) is always towards the supply and return/exhaust terminals and is indicated by *o*. The fitting total pressure loss of unequal inlet and outlet cross sections is calculated using the equation for fitting total pressure loss of equal inlet and outlet cross sections.

Psychometric Properties

When the temperature, relative humidity, and elevation are known, air density (ρ) and dynamic viscosity (μ) can be determined using the "show air properties" radio button in the output section of the DFDB (see Figure 1-12 for an example of the DFDB output showing the psychometric properties of standard air). Example 1-7 shows how to use the DFDB to obtain the psychometric properties for a specific location.

Duct Roughness Factors

Table 1-2 provides typical duct roughness factors. These must be supplied as input values when using the DFDB (ASHRAE 2016). Always consult the Duct Design chapter of the latest edition of *ASHRAE Handbook—Fundamentals* (2017a) for the most up-to-date information.

DFDB Features

Features of the DFDB include the following:

- Density and viscosity are calculated for any combination of elevation, temperature, and relative humidity.
- Pressure losses for rectangular, round, and flat oval ductwork are based on the Colebrook and Darcy equations (Equations 1-6 and 1-2).
- It includes user-selectable duct roughness factors.

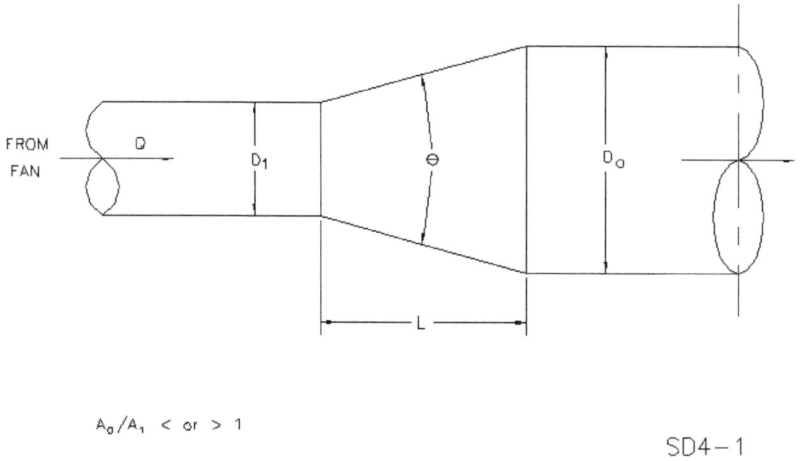

(a) Supply System Transition

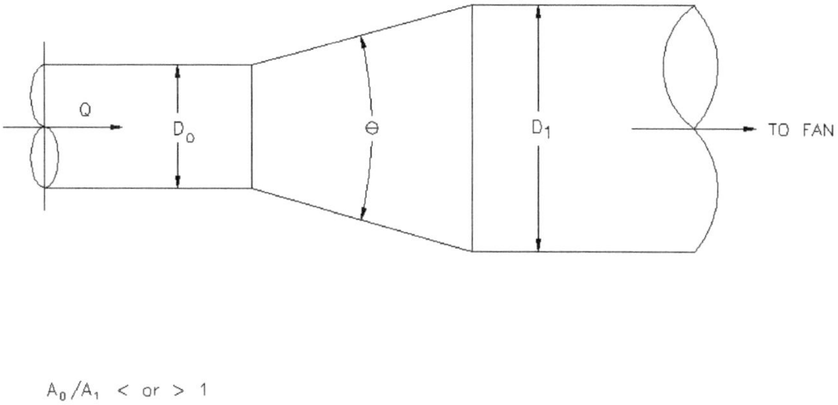

(b) Exhaust/Return System Transition

Figure 1-11 Examples of Transitions with Unequal Inlets and Outlets
(Reprinted from ASHRAE 2016, fittings SD4–1 and ED4–1)

AIR PROPERTIES		
Air Temperature	°F	70
Elevation	ft	0
Barometric Pressure	psia	14.969
Relative Humidity	%	0
Density (RHO)	lbm/ft^3	0.075
Viscosity (MU)	lbm/(ft-min)	0.000733500

Figure 1-12 DFDB Air Properties Output for Standard Air Conditions
(Reprinted from ASHRAE 2016)

Example 1-7.

Using the DFDB, determine the density and dynamic viscosity for Denver, Colorado, at 70°F and 50% rh.

Solution.

The elevation of Denver is approximately 5400 ft. The figure below shows the DFDB Air Properties output for the above inputs.

AIR PROPERTIES		
Air Temperature	°F	70
Elevation	ft	5400
Barometric Pressure	psia	12.046
Relative Humidity	%	50
Density (RHO)	lbm/ft^3	0.061
Viscosity (MU)	lbm/(ft-min)	0.000733500

DFDB Air Properties Output

(Reprinted from ASHRAE 2016)

Table 1-2 Duct Roughness Factors

(Reproduced from ASHRAE 2017a, Table 1)

Duct Type/Material	Absolute Roughness (ε), ft	
	Range	Roughness Category
• Drawn tubing	0.0000015	Smooth 0.0000015
• Polyvinyl chloride (PVC) plastic pipe • Commercial steel or wrought iron • Aluminum, round, longitudinal seams, crimped slip joints, 3 ft spacing	0.00003 to 0.00015 0.00015 0.00012 to 0.0002	Medium smooth 0.00015
Friction chart: • Galvanized steel, round, longitudinal seams, variable joints (Vanstone, drawband, welded. Primarily beaded coupling), 4 ft joint spacing • Galvanized steel, spiral seams, 10 ft joint spacing • Galvanized steel; spiral seam with 1, 2, and 3 ribs; beaded couplings; 12 ft joint spacing • Galvanized steel, rectangular, various type joints (Vanstone, drawband, welded. Beaded coupling), 4 ft spacing	0.00016 to 0.00032 0.0002 to 0.0004 0.00029 to 0.00038 0.00027 to 0.0005	Average 0.0003
Wright friction chart: • Galvanized steel, round, longitudinal seams, 2.5 ft joint spacing, ε=0.0005 ft	Retained for historical purposes	
• Flexible duct, nonmetallic and wire, fully extended • Galvanized steel, spiral, corrugated, beaded slip couplings, 10 ft spacing • Fibrous glass duct, rigid • Fibrous glass duct liner, air side with facing material	0.0003 to 0.003 0.0018 to 0.0030 — 0.005	Medium rough 0.003
• Fibrous glass duct liner, air-side spray coated • Flexible duct, metallic corrugated, fully extended • Concrete	0.015 0.004 to 0.007 0.001 to 0.01	Rough 0.01

- Calculations can be in either Inch-Pound (I-P) or Systéme International (SI) units.
- Fitting files can be copied, saved, and exported to Microsoft® Excel® or other similar programs.
- The DFDB is periodically electronically updated.
- Loss coefficient tables can be viewed and printed.
- Numbers and dates can be in any language convention.
- The DFDB interpolates the tables. Extrapolation is not allowed.
- Each fitting has a definite range of input values. For example, die-stamped elbows range from 3 to 10 in. and gored elbows range from 3 to 60 in.

NOMENCLATURE

A	= major axis of flat oval duct, in.
AR	= cross-sectional area of flat oval duct, in.^{2}
A_c	= cross-sectional area of common section, ft^2
A_s	= cross-sectional area of straight section, ft^2
A_b	= cross-sectional area of branch section, ft^2
A_d	= cross-sectional area of duct, ft^2
a	= minor axis of flat oval duct, in.
C	= fitting loss coefficient, dimensionless
c_o	= reference duct cross section
D	= diameter, in.
D_b	= branch section diameter, in.
D_c	= common section diameter, in.
D_e	= circular equivalent of rectangular or flat oval duct for equal length, fluid resistance, and airflow, in.
D_h	= hydraulic diameter, in.
D_s	= straight section diameter, in.
f	= friction factor, dimensionless
H	= length of one side of rectangular duct, in.
L	= duct length, ft
$\dot{m}$	= mass flow rate, lb_m/min
p	= pressure, in. of water
P	= perimeter of duct, in.
p_s	= static pressure, in. of water
P_s	= fan static pressure, in. of water
p_t	= total pressure, in. of water
P_t	= fan total pressure, in. of water
p_v	= velocity pressure, in. of water
Q	= airflow, cfm
Q_b	= branch volume flow rate, cfm
Q_c	= common (upstream/downstream) volume flow rate, cfm
Q_s	= straight-through volume flow rate, cfm
Re	= Reynolds number, dimensionless
V	= velocity, ft/min
W	= length of adjacent side of rectangular duct, in.

Symbols

$\Delta p_{dynamic}$	=	total dynamic pressure loss, in. of water
$\Delta p_{friction}$	=	friction pressure loss, in. of water
Δp_s	=	static pressure loss, in. of water
Δp_t	=	total pressure loss, in. of water
$\Delta p_{t,\text{ fitting}}$	=	fitting total pressure loss, in. of water
ε	=	absolute roughness, ft
μ	=	dynamic viscosity, $lb_m/(ft \cdot min)$
ρ	=	density, lb_m/ft^3

Subscripts

b	=	branch
c	=	common
o	=	reference duct cross-section
s	=	straight-through

REFERENCES

ASHRAE. 2009. Chapter 21, Duct design. In *ASHRAE handbook—Fundamentals*. Peachtree Corners, GA: ASHRAE.

ASHRAE. 2016. ASHRAE Duct Fitting Database, ver. 6.00.05 (online). Peachtree Corners, GA: ASHRAE. www.ashrae.org/technical-resources/bookstore/duct-fitting-database.

ASHRAE. 2017a. Chapter 21, Duct design. In *ASHRAE handbook—Fundamentals*. Peachtree Corners, GA: ASHRAE.

ASHRAE. 2017b. Chapter 3, Fluid flow. In *ASHRAE handbook—Fundamentals*. Peachtree Corners, GA: ASHRAE.

McGill. 2003. Chapter 1, Airflow fundamentals for supply duct systems. In *Duct System Design Guide*. Groveport, OH: McGill AirFlow Corporation.

Wright, D.K., Jr. 1945. A new friction chart for round ducts. *ASHVE transactions* 51:303–16.

2

Duct Design Considerations

OVERVIEW

The duct design methodologies presented in this book and found in Chapter 21, "Duct Design," of *ASHRAE Handbook—Fundamentals* (2017b) are equal friction, static regain, and constant velocity. These design methods are discussed in greater detail in this design guide in Chapters 3, 4, and 5, respectively. As a teaching tool, a manual design is presented for each method in this book, and blank templates and example problems in Microsoft® Excel® format are available at www.ashrae.org/DuctSyst. However, using the web-based ASHRAE Duct Fitting Database (DFDB; ASHRAE 2016) is recommended for obtaining the fitting total pressure loss coefficients, because they are not available in any other ASHRAE publications.

The equal friction method calculates each section's duct diameter from a friction chart (which is based on standard air conditions) or the DFDB (which can be configured for any air density). Based on a single friction rate, each duct section is sized to conform to that friction rate. The equal friction method can be semi-automated by using the DFDB to manually look up fitting loss coefficients and to calculate air density, velocity, velocity pressure, and the friction loss of straight duct.

Using the principles of the Bernoulli equation, where the total pressure is equal to the static pressure plus the velocity pressure, the static regain method calculates the duct diameter for each section so that the static regain, which is the decrease in velocity pressure for each section, is approximately equal to the total pressure loss for each section. The change in static pressure will be close to zero, which tends to produce a balanced design. This method allows higher design velocities to achieve the static regain without greatly increasing the operating pressure. This design method only applies to supply air systems. The static regain design method is usually performed by computer due to the number of repetitive iterations required to calculate the static regain in each duct section.

The constant velocity method is used for local exhaust systems. The design velocity in the duct system is generally set to a value such that particulate stays entrained in the air until it is collected. These velocities are generally higher than in the equal friction or static regain design methods, and the exhaust duct is usually under negative pressure.

Subsequent chapters cover other important topics that apply in general to any duct design method. For example:

- Suggestions are included for minimizing airborne equipment noise, duct breakout noise, flow-generated noise, and noise due to rooftop units and fan-powered air terminal units.
- A discussion is provided that addresses the effects of duct roughness.
- The weight of rectangular ducts is presented as a function of aspect ratio relative to circular ducts and the impact of fitting selection.

- Guidance is offered regarding where to locate balancing dampers in constant- and variable-flow systems.
- Tables are presented that show the maximum sizes of the following ducts that fit into clear spaces ranging from 18 to 38 in. in 4 in. increments:
  - A single round duct
  - Two round ducts in parallel
  - A rectangular duct with a 2:1 aspect ratio
  - A flat oval duct with an aspect ratio of 2:1

UNCERTAINTY AND ERROR CONSIDERATIONS

Measurement of the pressure losses in ducts and fittings is governed by the requirements in ANSI/ASHRAE Standard 120 (ASHRAE 2017a), which establishes uniform methods of laboratory testing of HVAC ducts and fittings to determine their resistance to airflow. The test results can be used to determine duct system flow losses in terms of pressure loss per unit length. Fitting losses are reported as local loss coefficients. The uncertainty range of the results is a measure of the potential error in a measurement or experimental result that expresses the confidence in the result at a specified level. Whenever testing is performed, the error analysis of the results should be provided by the test agency to give designers an indication of the expected accuracy of reported results. If the accuracy of the data used to design duct systems is very high (or the error in the data is very low), then the uncertainly level will be low and the designer should have confidence that the system will perform as designed.

Lower-pressure-loss types of fittings may have higher levels of uncertainty, because the accuracy of measuring low pressure losses could range up to 100%. For example, if an elbow had a stated pressure loss of 0.01 in. of water, the actual loss may be as much as 0.02 in. of water, i.e., the uncertainty range may be ±100%. This difference is inconsequential, however, since a pressure loss of 0.01 in. of water (or 0.02 in. of water) is well within the desired accuracy of the duct system pressure requirements. On the other hand, the uncertainty of high-pressure-loss fittings could affect the operating pressure and fan selection of the system, even if the pressure loss uncertainty is relatively low. If the pressure loss uncertainty of a fitting has a range of ±10% and the fitting has a presumed pressure loss of 1 in. of water, then the fitting could have an actual pressure loss as high as 1.1 in. of water. Using this type of high-pressure-loss fitting, such as a square elbow without turning vanes, in a duct system may require the fan to be sized larger to handle the uncertainty. What this is really saying is that the use of high-efficiency fittings minimizes the overall uncertainty in a duct system. If a system is designed for a fan total pressure of 2 in. of water, for example, and the total uncertainty is 0.1 in. of water, then the designer should be comfortable sizing the fan for 2.1 in. of water. However, if the overall uncertainty is 0.5 in. of water, then the fan total pressure should be selected for 2.5 in. of water.

Most duct and fitting losses determined using the measurement techniques from ASHRAE Standard 120 (2017a) will have uncertainties in the 10% to 15% range. However, this range was determined by tests conducted under ideal laboratory conditions. In real systems, close-coupled fittings and improper installations could result in much higher uncertainties. Close-coupled fittings could result in losses more than 50% higher than the sum of the individual losses. Poor installations can result in air being choked off and result in uncertainties greater than 100%. The use of close-coupled fittings should be minimized and accounted for in the system design. Installations should use good techniques to minimize unexpected pressure losses (such as a reduction to squeeze duct between two beams) caused by fittings that were not in the original design.

Remember, in duct design the term *design leg* refers to the particular pathway in the layout that exhibits the maximum pressure loss (i.e., the largest flow resistance). Because of uncertainties caused by experimental errors and poor installation, designers should not be concerned with path pressure losses that are within 10% of the critical path or design leg, as such designs are considered balanced, and balanced designs are desirable. Any path that has more than a 10% difference in pressure from the design path needs to have pressure losses added to force the proper amount of air to the desired terminal and achieve balance. Pressure losses can be increased by the addition of dampers, using smaller duct sizes, or using less-efficient fittings in the non-design legs.

SECTIONING A DUCT SYSTEM FOR DESIGN

Analysis and sizing of duct systems is based on a set of rules for numbering fittings and straight duct lengths together to form sections within a duct system. Every duct system is composed of lengths of straight duct and fittings. For the purpose of analysis, a new section is created whenever there is change in flow, size, or shape. For all other instances where the flow does not change and the entrance and exit dimensions are the same, the pressure loss associated with items such as constant-size elbows, dampers, fan system effect, and equipment is included in the same section.

The primary purpose of duct sectioning is to allow losses to be added in individual sections using total pressure loss coefficients for fittings and then multiplying the result by that section's velocity pressure. It is best not to use airflow direction when describing duct sections and instead use the section's position relative to the fan and supply and return/exhaust terminals. The general rule is that a fitting is assigned to the duct section that is closest to the terminal. In a supply air system the fitting is assigned to the downstream duct section, and in an exhaust/return air system the fitting is assigned to the upstream duct section. The main and branch coefficients for junctions are also referenced to the terminals. The rules for assigning and numbering duct sections are illustrated in Figure 2-1.

The starting point for the numbering of duct sections can be at an air-handling unit (AHU), fan, or terminal. The labeling should be sequential based on the starting point, either in the direction of flow or counter to flow. Because the numbering convention is different for supply and return systems, it is important to know that the supply air is the air from the positive side of the fan (outlet) that supplies the rooms or areas with the required air and the return air is air returned from the rooms or areas to the inlet side of the fan. Following are examples of starting points and numbering sequences:

- *Starting at the AHU.* First number the supply duct system toward the supply air terminals, then proceed with the return duct system toward the return air terminals. This is the convention used in Figure 2-1.
- *Starting at the terminals.* First number the return duct system toward the AHU, then proceed from the farthest supply air terminal toward the AHU.
- *Direction of airflow.* First number the supply duct system toward the supply air terminals, then proceed with the return duct system from the return air terminals toward the AHU.

Sectioning requires identification by numbers or tags. Small systems are simple to identify—they can use 1-2-3-4, etc., as shown in Figure 2-1. Larger systems with many sections need a logical assignment of numbers. It is best to work out a sequence of numbers by mains, branches, and terminals that also represents the floors in multistory systems using a three- or four-digit sequence. For example, first-floor ductwork sections might be tagged 0101-0102-0103, etc., and second-floor ductwork sections might be tagged 0201-0202-0203, etc. Final terminal sections need higher tags such as 1001-1002-1003, etc., for the first floor and 2001-2002-2003, etc., for the second-floor terminations.

Example 2-1 shows one way of assigning section numbers for a supply system.

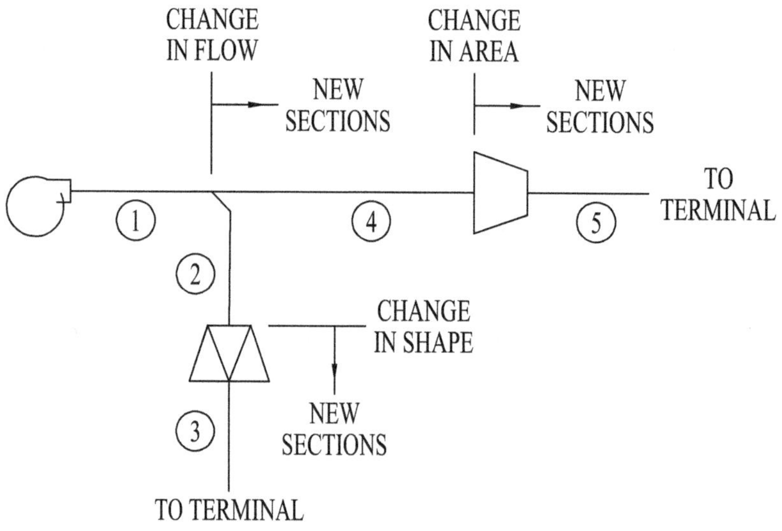

(a) Supply System

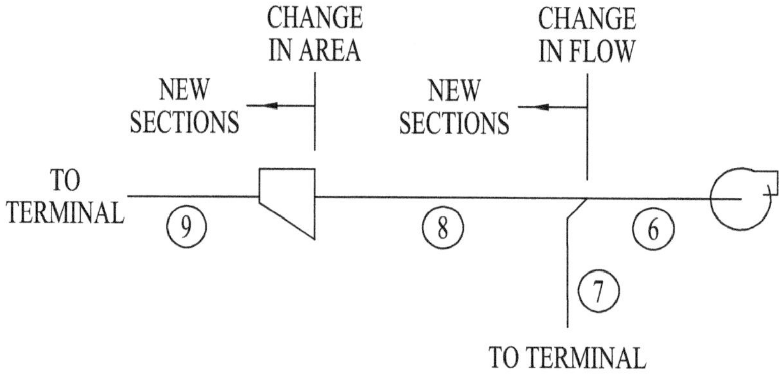

(b) Return/Exhaust System

Figure 2-1 Examples of How Duct Sections are Assigned and Numbered

SELECTING A DUCT DESIGN METHOD

Table 2-1 describes the equal friction, static regain, and constant velocity duct design methods. Study of this table will provide insight into selecting the appropriate duct design method. The equal friction method can be used to design supply, return, or exhaust systems; it is discussed in more detail in Chapter 3. The static regain method is only applicable to supply systems and is discussed in more detail in Chapter 4. The constant velocity method is a more specialized approach used to design local industrial exhaust systems and is discussed in greater detail in Chapter 5.

Comparing the Equal Friction Design Method and the Static Regain Design Method

The equal friction design method (refer to Chapter 3 for a design example) sizes duct sections based on a constant friction rate. The static regain method (refer to Chapter 4 for a design example) designs duct sections so the change in static pressure from junction to junction is almost zero, resulting in a balanced system.

Example 2-1.
For the supply system shown in the following figure, determine the sections and number them according to the rules. Start at the fan.

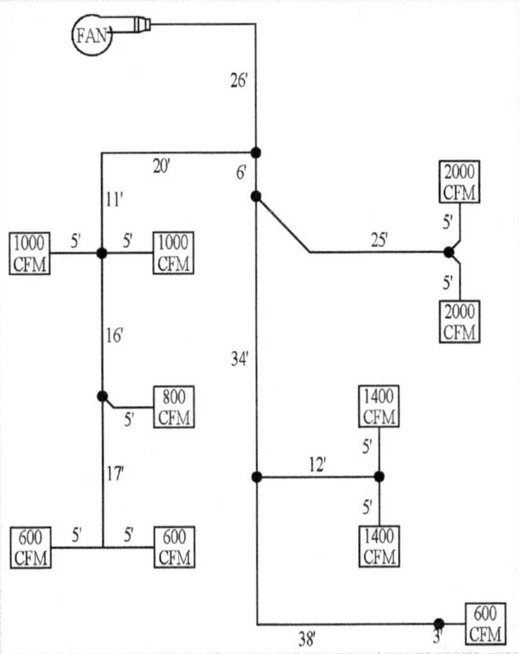

Sample System

Solution.
The solution is shown below. Note that all junction fittings are assigned to the section downstream or toward the terminals.

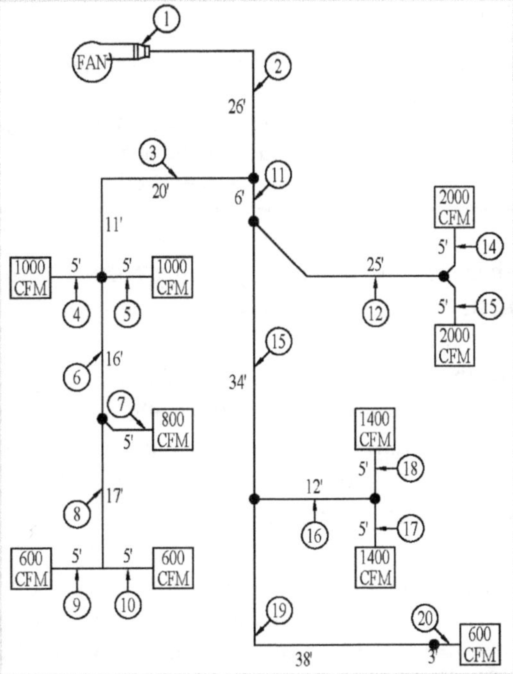

Section Number Assignment

Table 2-1 Duct Design Methods

Design Method	Duct Sizing	Notes
Equal friction	A section's duct size is based on the chosen friction loss rate (in. of water per 100 ft of duct) for all duct sections, or the duct size at the fan is selected based on the maximum air velocity so that aerodynamic noise is not a problem. Then the friction rate for that section is used to size all the other sections.	Feasible when done semiautomatically, using a spreadsheet such as those referenced in Chapter 3 or the DFDB (ASHRAE 2016) to get fitting loss coefficients, as the DFDB also calculates density, velocity, velocity pressure, and the friction loss of straight duct. Can be used with round, flat oval, or rectangular ductwork. It is recommended that equal friction designs be balanced. This requires resistance being added to non-critical paths with dampers, smaller duct sizes, or less-efficient fittings being selected. Calculations may be tedious and time consuming. Systems will be easier to balance if the velocity is limited to 1500 fpm. Can be used to design supply, return, or exhaust systems.
Static regain	The duct size at the fan is selected based on the maximum air velocity so that aerodynamic noise is not a problem. Other main ducts are sized to achieve static regain from section to section. The branches are also sized by static regain using the entering junction velocity as the starting point. A minimum static pressure is maintained at the terminal sections so that sufficient energy is available for proper operation of variable-air-volume (VAV) air terminal units or constant-velocity (airflow) diffusers.	For use with computer programs linked to the DFDB (ASHRAE 2016). An iterative solution is required to determine the duct size of each main section. Can be used with round, flat oval, or rectangular ductwork. This method is self-balancing because the change in total pressure is equal to the change in velocity pressure, which causes the change in static pressure to be zero. Use of this method on low-pressure systems is limited. Only applicable to supply systems.
Constant velocity	The duct size is based upon maintaining a constant-velocity exhaust system upstream of particle collectors.	Applications are exhaust systems conveying fumes or particulates. Balance is obtained by using gate dampers in non-particulate airstreams and increasing flow rates in particulate-conveying systems where additional resistance is required.

When comparing duct design methods, operating costs and first costs should be compared. If the total pressures of both designs are about equal, then whichever duct design method gives the lower first cost is most economical.

In the comparison of the two duct design methods for supply-air systems (see Example 3-1 for the equal friction method and Example 4-1 for the static regain method), the equal friction method has eight duct sections with a smaller diameter than the duct in the static regain method (Table 2-2), so all else being equal for this example, the equal friction method has a lower first cost. However, these smaller sizes result in a larger pressure drop, so the total pressure requirement for the equal friction method is 12% higher than that of the static regain method, leading to higher operating cost (fan energy). Higher operating cost generally results in higher total costs (first cost plus operating costs, as the operating costs recur). For either design approach, significant reductions in the weight of the ductwork can be gained by using lighter gauges with round spiral lockseam construction.

In addition to first cost and operating cost, the system imbalance should be considered. Systems should be designed to have minimal excess pressure for any path so that the airflow is prop-

Table 2-2 Comparison of Duct Sizes and Pressure Losses

Duct Section	Equal Friction Design		Static Regain Design	
	Size, in.	Δp_t, in. water	Size, in.	Δp_t, in. of water
1	H_1 = 20, W_1 = 27	Included with section 2	H_1 = 20, W_1 = 27	Included with section 2
2	25	0.19	25	0.19
3	17	0.42	20	0.37
4 and 5 (outlets)	10	0.67	10	0.61
6	13	0.09	16	0.04
7 (outlet)	9	0.57	9	0.53
8	11	0.08	14	0.03
9 and 10 (outlets)	8	0.52	8	0.48
11	21	0.12	22	0.10
12	17	0.33	21	0.26
13 and 14 (outlets)	14	0.61	14	0.54
15	16	0.13	17	0.10
16	15	0.15	17	0.11
17 and 18 (outlets)	12	0.16	12	0.56
19	8	0.25	9	0.15
20 (outlet)	8	0.50	8	0.49

Note: Outlets were presized for the air terminal unit.

erly distributed. If not, excess restrictions (by the use of dampers) may be necessary. Using smaller duct sizes and less-efficient fittings to help balance the duct design is what is referred to as *total pressure design*.

Examples 3-1 and 4-1 show that both systems are fairly well balanced. The static regain design shows how using a smaller section diameter helps create even less excess pressure (path J). For the equal friction design, a constant friction rate of 0.41 in. of water per 100 ft was calculated. This design was reasonably balanced because the system is somewhat symmetrical. The static regain example was in balance by design (Table 2-3).

Both examples had an acoustical requirement of Room Criterion (RC) 35 in the first section, which leads to a maximum velocity of 3500 fpm using round duct. Rectangular designs have much lower velocity limit requirements, so they may also have much higher first costs. A rectangular RC35 velocity limit would be 2500 fpm, resulting in a 40% larger duct cross-sectional area for rectangular duct.

SOFTWARE

The primary benefit of duct design software is that the tedious activities of looking up loss coefficients and solving iterative calculations are accomplished automatically. As a result, the labor to

Table 2-3 Comparison of Total Pressure Requirement and Balancing

Path	Equal Friction Design (from Example 3-1)		Static Regain Design (from Example 4-1)	
	Total Pressure, in. of water	Excess Pressure, in. of water	Total Pressure, in. of water	Excess Pressure, in. of water
A/B (4/5)	1.28	0.02	1.16	0.00
C (7)	1.27	0.03	1.12	0.04
D/E (9/10)	1.30	0.00	1.09	0.07
F/G (13/14)	1.26	0.04	1.08	0.08
H/I (17/18)	1.20	0.09	1.06	0.11
J (20)	1.19	0.11	1.04	0.13

design (or redesign) duct systems is not as intensive, which results in economically designed ductwork for the owner. Minimum requirements for good duct design software are as follows:

- The program should interface with the current version of the DFDB (ASHRAE 2016).
- The program should perform a complete acoustical analysis that is in compliance with Chapter 8 in this guide or the latest Noise and Vibration Control chapter of *ASHRAE Handbook—HVAC Applications* (2019).
- The program should be validated against the example problem in the most current Duct Design chapter of *ASHRAE Handbook—Fundamentals* (2017b) and the example problems included in this design guide.

Regardless of the software program used, it is essential that time be taken to learn the program, understand the program algorithms, and recognize the program limitations.

DUCT DESIGN CONSIDERATIONS

Acoustics—Noise and Vibration Control

Acoustics is an important part of duct design (it is discussed in greater detail in Chapter 8). Problems that can be encountered when acoustics is ignored are

- airborne equipment noise,
- equipment vibration,
- duct-borne fan noise,
- duct breakout noise,
- duct break-in noise
- flow-generated noise, and
- duct-borne cross talk.

Sources of noise generated at undesirable levels that may need to be treated can be the following:

- rooftop units,
- air terminal units,
- return air openings,
- outlets, and
- duct fittings.

Each of these problems and sources is discussed briefly in this section.

Airborne Equipment Noise

Airborne equipment noise is noise generated by the HVAC system that can be heard above the background noise level in a space, such as a room or hallway. The following guidelines are offered to reduce airborne equipment noise:

- Do not place a mechanical penthouse on a lightweight bar joist roof over a noise-sensitive area.
- Use noncritical spaces to buffer critical spaces from equipment room noise. These buffer spaces include corridors, elevator cores, toilet rooms, storage rooms, telephone equipment rooms, and other similar spaces.
- Equipment room floors and ceilings must provide adequate sound insulation to protect adjacent spaces from noise transmission.

Equipment Vibration

Equipment vibration can be created by poor installation, fan blade imbalances, or other reasons. To avoid problems caused by equipment vibration, do not place AHUs close to walls or ceilings. There is a phenomenon called *close coupling* in which a small air space will conduct cabinet vibratory motion to the wall or ceiling. A space of approximately 3 ft usually suffices. Provide a nominal 4 in. thick housekeeping pad beneath equipment cabinets to minimize the effects of close coupling to the floor.

Like airborne equipment noise, the principal strategy for controlling equipment vibration is location. Put the equipment on grade, in the basement, or in the sub-basement (the exceptions are fans and AHUs, unless outdoor air is readily accessible by this equipment in these spaces). Do not place equipment on long-span floors or roofs. The structure must be at least ten times as stiff as the equipment isolators. Do not use housed isolators or thin pads to restrict vibratory motion.

Duct-Borne Fan Noise

Duct-borne fan noise is noise generated by the fan that propagates to conditioned spaces via the ducts. The best way to control duct-borne fan noise is to not create it. Select quiet fans based on sound power data (see Chapter 8 for a definition of sound power). Do not specify noisy fans and try to “fix” them. Provide good fan outlet conditions by eliminating poor fan connections to ductwork (see Figure 7-7 in Chapter 7). Do not turn the air in the “wrong” direction or the ducts will rumble.

Duct Breakout Noise

Duct breakout noise is noise that escapes through the duct walls. It is often caused by noise from the fan. To combat duct breakout noise, provide in-duct silencers or long runs (15–20 ft) of internally lined duct to reduce fan noise prior to a duct running above the acoustical tile ceiling of an occupied space. Acoustical tile ceilings provide almost no attenuation of low-frequency noise. When running duct in an acoustical tile ceiling, use 1 in. thick duct liner instead of 0.5 in., which is too thin to be useful in these situations. The 1 in. thick acoustical duct is effective for higher frequencies, but low-frequency rumble from fans or generated noise requires at least 2 in. of liner. For critical noise areas, the connection of VAV air terminal units to the branch ductwork should be made with metal round duct, not flex duct. Also, make sure that there are at least three duct diameters of straight duct between VAV air terminal units and any elbows or reducers.

Duct Break-In Noise

Duct break-in noise is noise introduced to ductwork through the duct wall in one location (for example, an area with high noise levels) that propagates throughout the rest of the duct system. If break-in noise is a problem, thicker duct walls or other method of increasing the transmission loss from the surrounding to the duct must be employed.

Flow-Generated Noise

Flow-generated noise is sound generated by improper flow conditions at higher velocities or by dampers. To avoid problems from this, select duct fittings for smooth flow and gradual velocity changes. Chapters 3 and 4 include tables of maximum air velocities that can aid in avoiding generated noise under various installation conditions.

Constant-volume systems require balancing dampers; VAV systems do not require dampers—see the Constant-Volume (CV) Systems and Variable-Air-Volume (VAV) Systems subsections in the Balancing Dampers section of this chapter for more information on dampers. In noise-sensitive applications, locate constant-flow dampers well upstream of the outlet within lined ductwork which give the noise they can generate time to attenuate before reaching the diffusers (see Table 8-18 in Chapter 8 for recommended velocities at diffusers).

Duct-Borne Cross Talk

Duct-borne cross talk occurs when the sound of voices or other noises propagate through the ductwork from one space to another. This can be enabled by short runs of ductwork, which can cross-connect rooms, the duct working like a speaking tube. To minimize duct-borne cross talk, provide adequate runs of internally lined ductwork, with elbows. Dumping return air into a ceiling plenum above a common corridor can also cause cross talk unless adequate return ductwork or silencers are provided.

Rooftop Units (RTUs)

Rooftop AHUs cause more than their share of noise problems. Locate RTUs with extreme care over toilet rooms, storage rooms, lunch rooms, or other noncritical spaces. Do not place RTUs on long-span roofs. If the roof is not stiff at the mounting location, provide a structural steel frame to transfer the weight to bearing walls, columns, or use structural curbs with rails extended to spread the weight.

Air Terminal Units

- To eliminate radiated sound from VAV air terminals, locate them above noncritical areas such as corridors, copy machine areas, and work areas. Mounting air terminals over noise-sensitive areas should be avoided.
- Where possible, locate air terminals in the largest possible ceiling plenum volume. Larger plenums generally increase the ceiling space effect. Good practice dictates that at least 2 in. clearance be established between the ceiling tile and the bottom of the unit.
- Sound generated by air terminal dampers increases as a function of both inlet airflow velocity and inlet static pressure. Try to design duct systems that provide adequate but not excessive static pressure at the air terminal primary air inlet. Upsize inlet selections to gain lower inlet velocities and quieter terminal acoustical ratings.
- Terminal units that are exposed in occupied areas without ceilings are common sources of noise complaints. Ceilings are almost always required to control terminal unit radiated noise.
- For hospitals, the rating of terminal units should not include fiberglass-lined duct or flexible duct, because they are not allowed. For these cases the change in sound power should be calculated by use of the methods provided in Chapter 8, because manufacturers' published noise criteria (NC) data include 5 ft of 1/2 in. lined duct and 5 ft of flex duct. The

difference between published values and values for VAV terminal units without lined duct and no flex duct is significant.

- VAV air terminal unit inlet connections should be constructed from sheet metal; do not use flexible duct. Sheet metal ensures a smooth inlet to the VAV terminal unit velocity pressure sensor, thereby improving airflow measurement accuracy. It also reduces breakout noise from the VAV terminal unit damper, since flex duct is virtually transparent to noise.

Return Air Openings

Return air openings provide a direct sound path through a ceiling. Avoid locating unducted returns directly below air terminals and adjacent flexible ducts.

Outlets

Manufacturers' diffuser data assume 10 diameters of straight duct upstream and 10 dB room absorption, neither of which is realistic. It is therefore recommended that designers add 5 NC to published values for ceiling diffusers and 3 NC for slot diffusers (straight inlet: no 90° elbow at inlet). For diffuser installation instructions consult Figure 8-15 in Chapter 8. Use limited-length flexible duct upstream installed in compliance with the Air Diffusion Council's installation standards (ADC 2010), which require the flexible duct to be fully extended with no sag. Because of its thin walls, flexible duct installed to the diffuser allows duct-borne noise to break out above the ceiling when.

Duct Fittings

Generated sound in ducts occurs due to fittings with abrupt transitions, sharp edges, and adjacency to other fittings, especially at higher velocities. Avoiding these circumstances by design can prevent excessive generated sound.

PRESSURE LOSS CONSIDERATIONS OF DUCTS AND FITTINGS

Duct roughness, shape, and size as well as the number of fittings and their types, shapes, and sizes can all have a profound effect on duct system pressure losses. It is important to minimize the pressure losses in the design legs. Note, however, that if the duct section being analyzed for pressure loss is not in the design leg, the pressure loss will have no effect on the operating pressure of the duct system.

Impact of Duct Roughness

The impact of duct material roughness factors is shown in Figure 2-2, where the duct roughness categories from Chapter 1 are plotted for a 10 in. diameter round duct. The plotted data were obtained using the DFDB (ASHRAE 2016). Designers should consider the effect of material roughness on operating costs when designing systems with materials of high resistance.

Flexible duct, fully extended, is categorized as "medium rough." The effect of installed, compressed (not fully extended) straight flexible duct is shown in Figure 2-3; from this the designer can determine a pressure drop correction factor (PDCF) that gets applied to the pressure drop for friction loss of a medium rough duct (with $\varepsilon = 0.003$ ft). The relationships relative to the other roughness categories for a 10 in. diameter flexible duct installed 4% and 10% compressed are illustrated in Figure 2-2. Flexible duct should be installed in a fully stretched condition (or as close as possible), but never at more than 4% compression. One challenge that the industry faces is that improperly installed flexible duct is difficult to detect, because compression is difficult to see or measure when installed (Culp 2011). It is also best practice to minimize bends and avoid kinks. Abushakra et al. (2004) show that loss coefficients for bends in flexible ductwork vary widely from

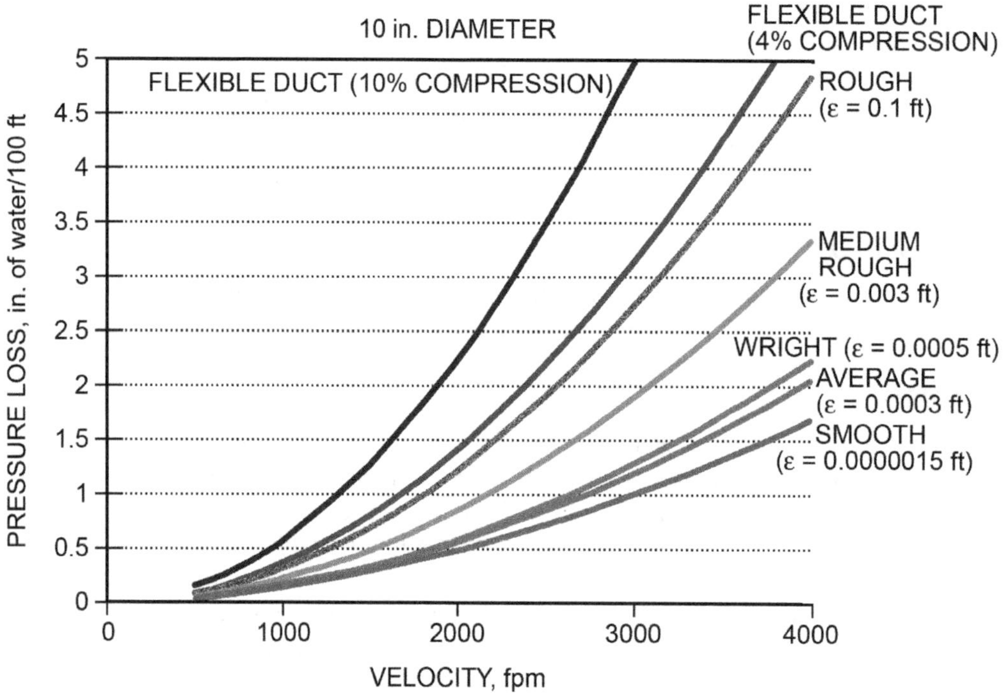

Figure 2-2 Duct Material Surface Roughness Impact on Pressure Loss
(Adapted from ASHRAE 2017b, Figure 11)

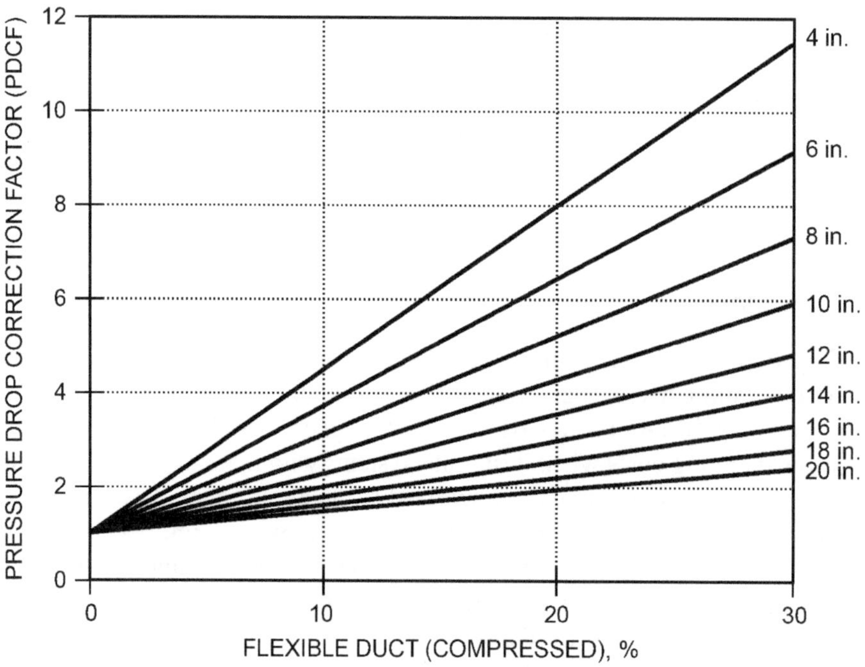

Pressure drop correction factor (PDCF) referenced to ε = 0.003 ft (medium rough) (Abushakra 2004; Culp 2011)

Figure 2-3 Pressure Drop Correction Factor for Unextended Flexible Duct
(Reprinted from ASHRAE 2017b, Figure 8)

condition to condition, with no uniform or consistent trends. Loss coefficients vary from a low of 0.9 to a high of 3.3. For comparison purposes, for an 8 in. die-stamped 90° elbow with a centerline *r*/*D* ratio of 1.5, the loss coefficient is 0.11.

Impact of Fittings

Dynamic losses result from flow disturbances, including changes in direction associated with tees/laterals (wyes), transitions, and duct-mounted equipment. The loss coefficients for comparable round and rectangular ducts are roughly the same. It is best to use reasonably low-pressure loss fittings such as radius elbows and avoid high-pressure loss fittings such as mitered elbows without vanes. Designers also should not lay out duct systems with less than five hydraulic diameters between fittings where possible, because close-coupled fittings can dramatically increase pressure drop. Fitting loss coefficients in the DFDB (ASHRAE 2016) are based on fully developed flow. If flow is not fully developed because of length limitations, the pressure losses will be higher than the DFDB calculations.

Fan Connections

Designs should avoid poor connections to centrifugal and axial fans. Designers with access to the DFDB should review fitting series SR7, SD7, ER7, and ED7 to minimize the fan system effect at the outlets or inlets of fans (ASHRAE 2016; refer to Table 1-1 in Chapter 1 for an explanation of the DFDB nomenclature).

Elbows

Use radius elbows (Figure 2-4) with rectangular duct rather than mitered elbows with turning vanes (Figure 2-5) whenever space permits. On medium- and high-velocity VAV systems (>2000 fpm), where a full-radius elbow cannot fit, a short-radius elbow with one or more splitters (Figure 2-4c) should be used. For short-radius elbows with splitter vanes, provide sufficient duct length upstream and downstream so that flow is fully developed. For round or flat oval elbows use $r/D = 1.5$ where possible and $r/D = 1.0$ if space is limited. Avoid mitered elbows unless absolutely necessary. If required, use mitered elbows with turning vanes. Turning vanes should only be used on low-velocity (2000 fpm maximum) systems where radius elbows will not fit.

Tees

Conical taps or 45° entry taps (Figures 2-6 and 2-7) are generally the most efficient branch fittings and should be used in the design legs of duct systems, with the 45° entry taps such as fitting SD5-12 being slightly more efficient (Figure 2-6b). If designers are not sure which path will be the design path, then the taps should be used at all 90° branch fittings. Wyes are typically used for converging flow in industrial exhaust systems. Straight (spin-in) branches of tees should be limited to sections of ductwork where the velocity does not exceed 2000 fpm because they have higher loss coefficients. Figures 2-6 and 2-7 show some common 90° branch fittings—diverging fittings on the left column and converging fittings on the right. Figure 2-8 shows some fittings that can be used at less than 2000 fpm (but should not be used if the velocity exceeds 2000 fpm).

Duct Shape Selection

No Space Constraints

Round ductwork is generally the preferred shape when adequate space is available, for the following reasons:

- The weight of round ductwork is lower than that of equivalent-sized rectangular or flat oval ducts. Figure 2-9 shows the relative weights of rectangular duct to round duct for duct pres-

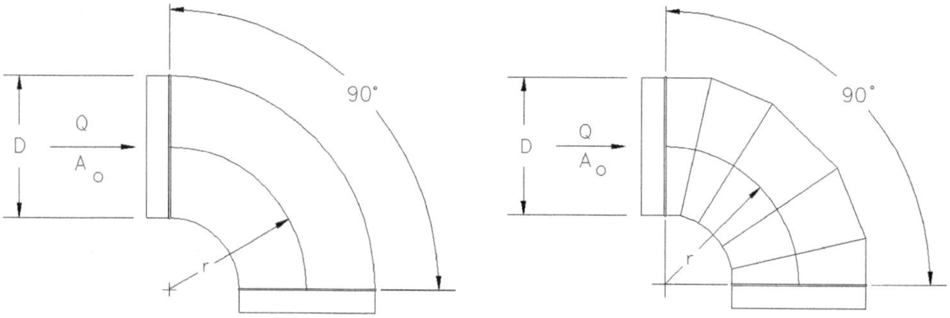

(a) Round Elbow, Radius/5-Gore (fittings CD3-1 and CD3-9)

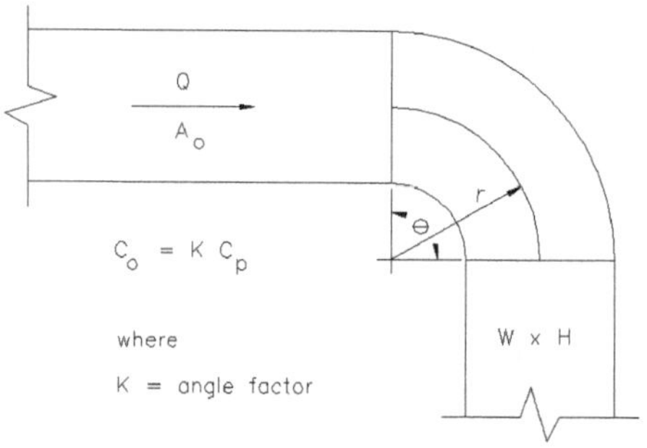

(b) Rectangular Elbow, Radius (fitting CR3-1)

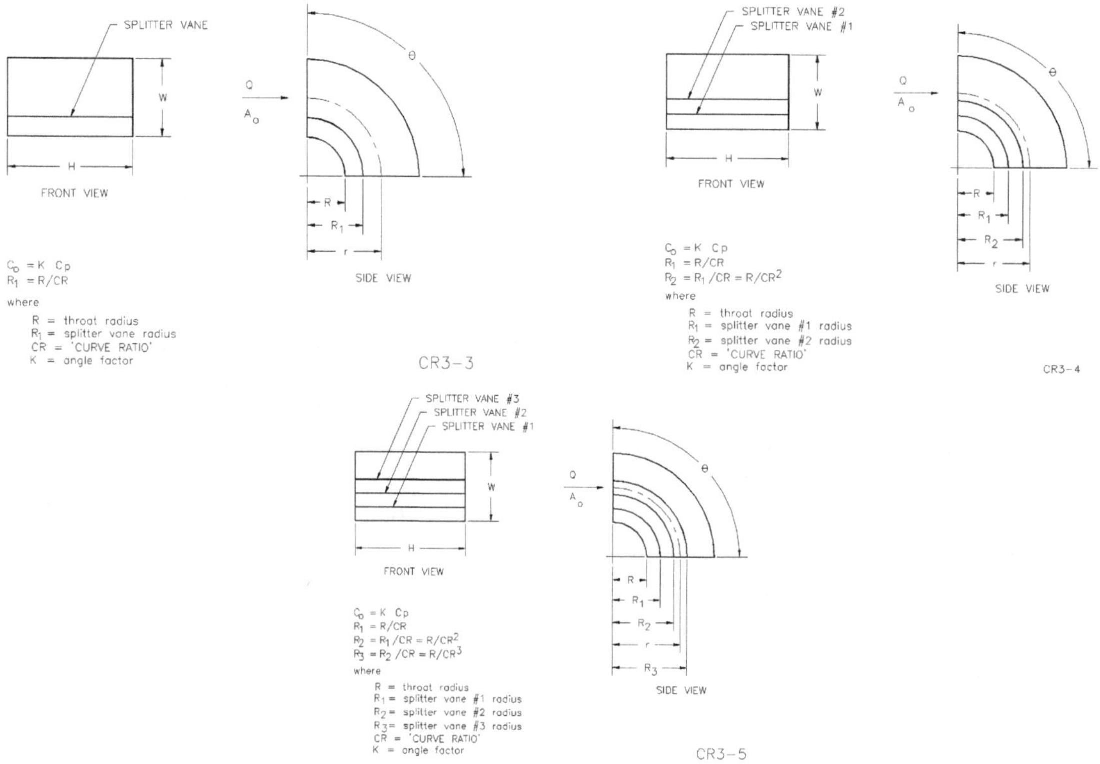

(c) Rectangular Elbow, Radius with 1, 2, or 3 Splitter Vanes (fittings CR3-3, CR3-4, and CR3-5)

Figure 2-4 Radius Elbows
(Reprinted from ASHRAE 2016)

sures from ±0.5 to ±10 in. of water when the equivalent diameter of the rectangular duct is the same as the round duct diameter. *Equivalent diameter* is defined as the effective diameter of a duct that is not round but that has equal resistance to flow for equal flow and length.

- The perimeter of round ducts is less than that of equivalent rectangular ducts. For rectangular ducts with aspect ratios from 2 to 4, the increase is approximately 30% to 55%. This increase results in increased insulation.
- Round ducts have an excellent resistance to low-frequency breakout noise (Schaffer 2005).

Space Constraints

Space limitations and obstructions (particularly ceiling height) frequently cause problems with the installation of ductwork. In these cases, the choices are multiple runs of either round, rectangu-

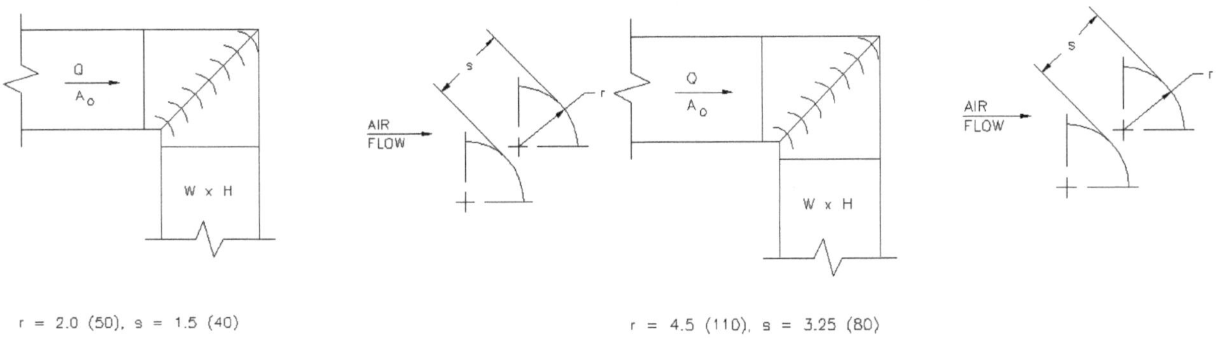

(a) Single-Thickness Vanes (fittings CR3-9 and CR3-12)

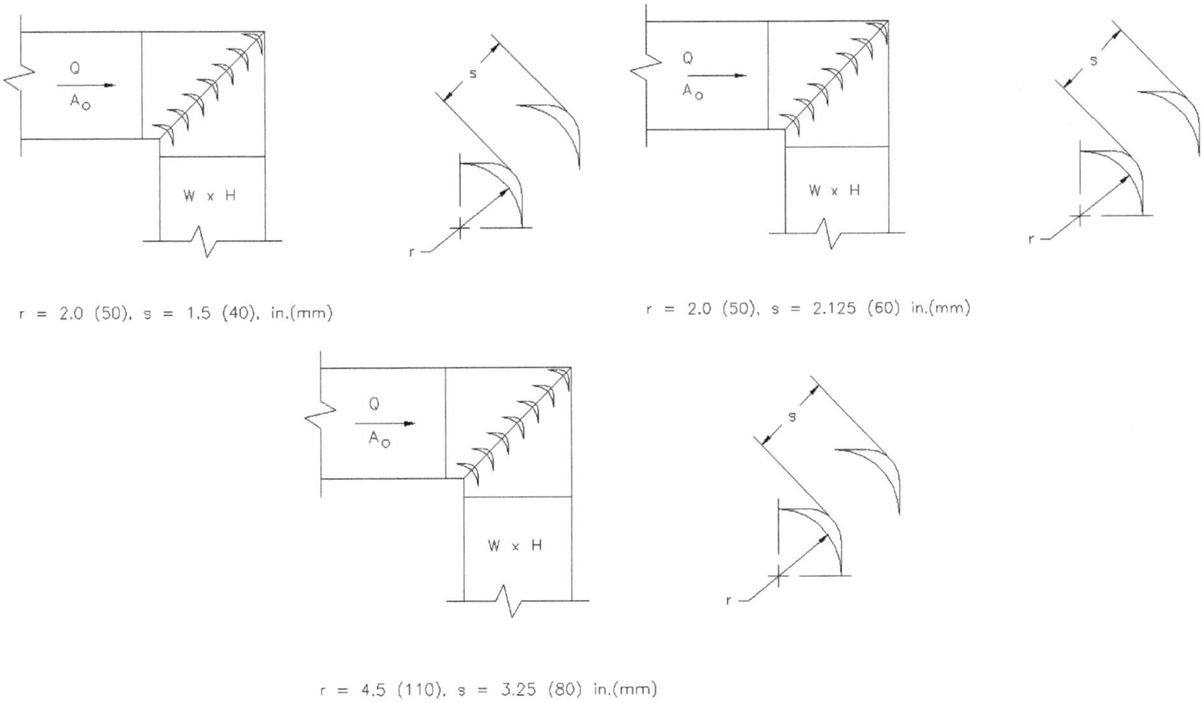

(b) Double-Thickness Vanes (fittings CR3-14, CR3-15, and CR3-16)

Figure 2-5 Mitered Rectangular Elbows with Turning Vanes
(Reprinted from ASHRAE 2016)

lar, or flat oval ducts, depending on the air quantity that needs to be conveyed by the duct. Table 9 in Chapter 21, "Duct Design," of *ASHRAE Handbook—Fundamentals* (2017b) address three design cases: 0.08, 0.2 and 0.6 in. of water per 100 ft friction rates. If the air quantity exceeds the upper value of the rectangular duct, the duct layout should be reconfigured. When selecting rectangular or flat oval duct, consider the following:

- Low-frequency breakout noise for flat oval duct is good but for rectangular duct is fair (Schaffer 2005). As a form of noise control, there may be situations where designers want the noise to break out when the duct is on a roof and not near a critical area where the noise could create a disturbance.
- The use of rectangular duct instead of flat oval (when sized for equivalent diameter) increases the perimeter roughly 17% to 70% for aspect ratios ranging from 1 to 4, respectively. This has the effect of increased surface area.

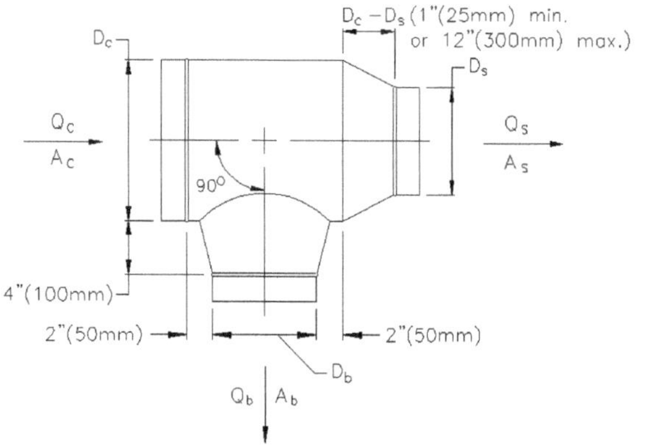

Conical Tee, Diverging (fitting SD5-10)

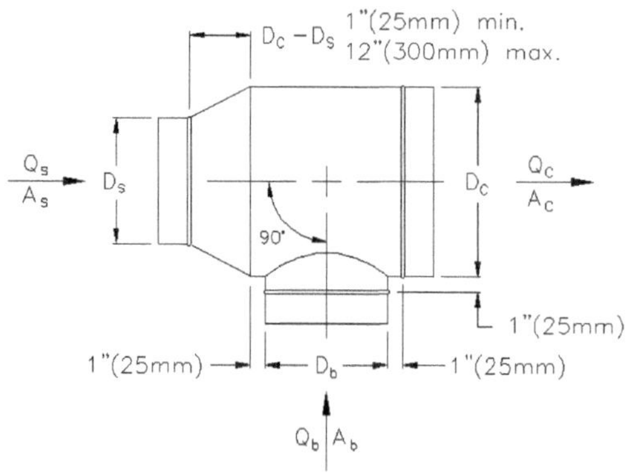

Conical Tee, Converging (not available in DFDB, use fitting ED5-3 [shown] instead)

(a) Tee, Round, Conical Takeoff

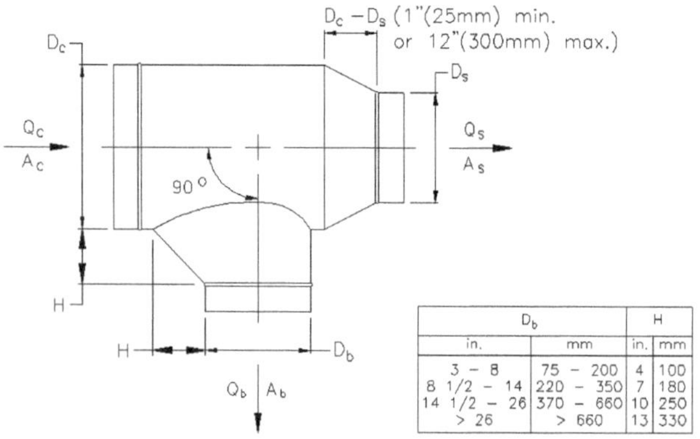

D_b		H	
in.	mm	in.	mm
3 - 8	75 - 200	4	100
8 1/2 - 14	220 - 350	7	180
14 1/2 - 26	370 - 660	10	250
> 26	> 660	13	330

45° Entry Tee, Diverging (fitting SD5-12)

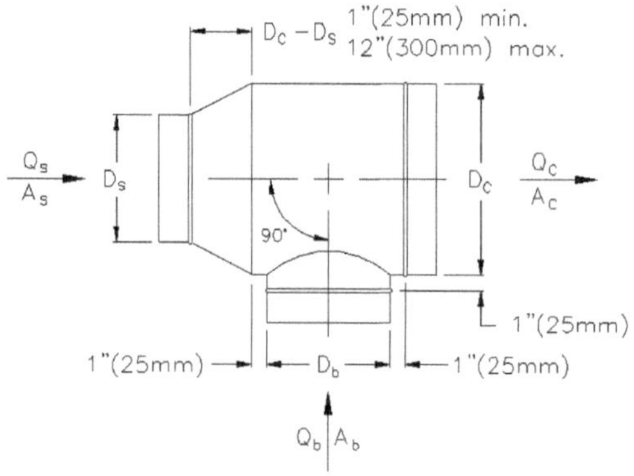

45° Entry Tee, Converging (not available in DFDB, use fitting ED5-3 [shown] instead)

(b) Tee, Round, 45° Entry Tap

Figure 2-6 90° Branch Fittings, Round
(Reprinted from ASHRAE 2016)

- Rectangular duct is typically available in 4, 5, or 6 ft lengths. Spiral round and flat oval duct can be provided in longer lengths. However, smaller lengths of rectangular duct may be easier to handle and install, especially for where large duct sizes are required.
- Certain sizes of spiral flat oval duct may not be available from all manufacturers due to limitations of machine capability, availability of a particular mandrel size, or their ability to make odd diameters of duct. Designers should consult manufacturers for available dimensions.

Duct Design with Other Duct Shapes

Round duct systems are the most efficient from weight, cost, pressure, and energy-loss standpoints. Flat oval duct is closer to round than is rectangular and therefore has many of the benefits of round duct, along with some of the characteristics of rectangular duct. Whatever shape is finally used, the ductwork is typically designed assuming round duct and then converted to rectangular or flat oval duct using Equations 1-7 or 1-8 of Chapter 1. However, when calculating the actual pressure losses the velocity should be based on the final duct cross section (flat oval or rectangular), not the velocity that could be calculated using the equivalent round sizes. Likewise, the appropriate

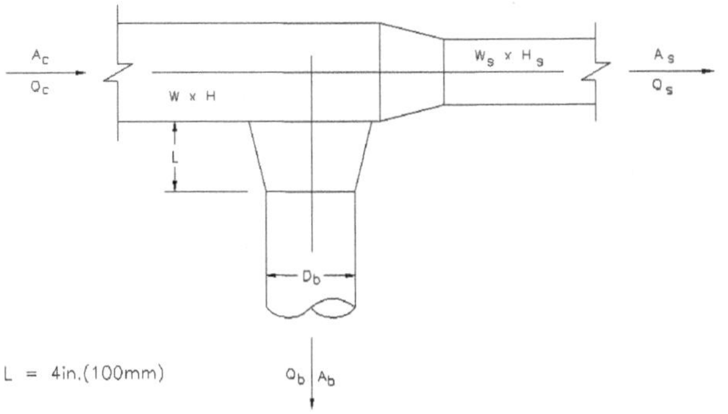

Conical Tap, Diverging (fitting SR5-12)

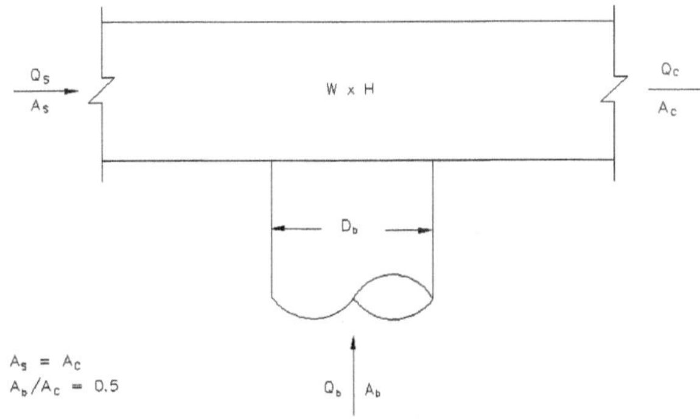

Conical Tap, Converging (not available in DFDB, use fitting ER5-2 [shown] instead)

(c) Tee, Rectangular, Conical Takeoff

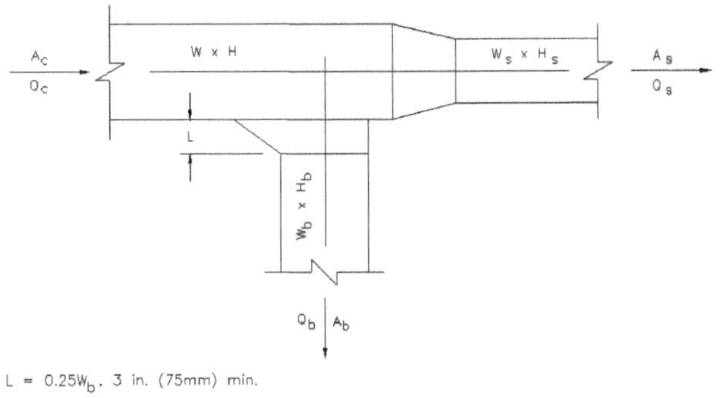

Diverging (fitting SR5-13)

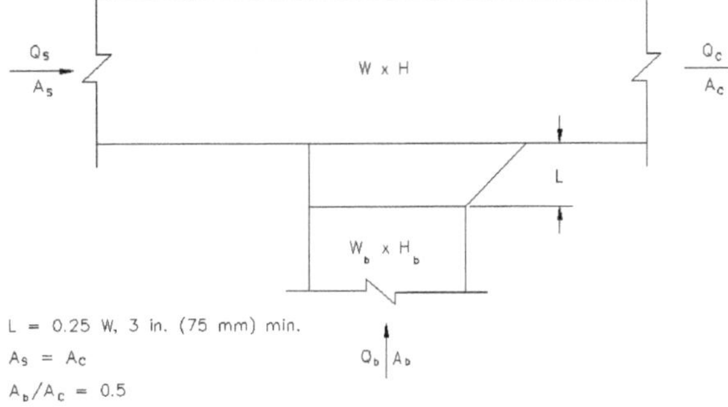

Converging (fitting ER5-3)

(d) Tee, Rectangular, 45° Tap

Figure 2-7 90° Branch Fittings, Rectangular
(Reprinted from ASHRAE 2016)

hydraulic diameter (for the round, rectangular, or flat oval duct) should also be used to calculate the friction factor in the Colebrook equation (Equation 1-6).

Duct Material Selection

Refer to Chapter 7 for explanations of common materials used when fabricating ducts (G-60, G-90, aluminum, stainless steel, and others). No matter the material chosen, it is recommended that flexible duct not be allowed except for 6 ft upstream of diffusers and light troffers. Flexible duct, if allowed, must meet accepted industry installation practices (ADC 2010). As a minimum, the sheet metal worker must install flexible duct extended with no more than 4% compression. Do not install the duct in its compressed state or use excess length. Suggested best practice is to remove the compressed ducts from their packaging then apply sufficient axial force needed to return the duct to its full length. After that, allow the flexible duct to relax while resting on the floor, then cut it to the desired length. Allow 2 in. on either end of the duct to accommodate overlap of the flexible duct with steel fittings for connecting draw bands. To a close approximation, this technique yields a flexible duct compression that does not exceed 4% upon installation.

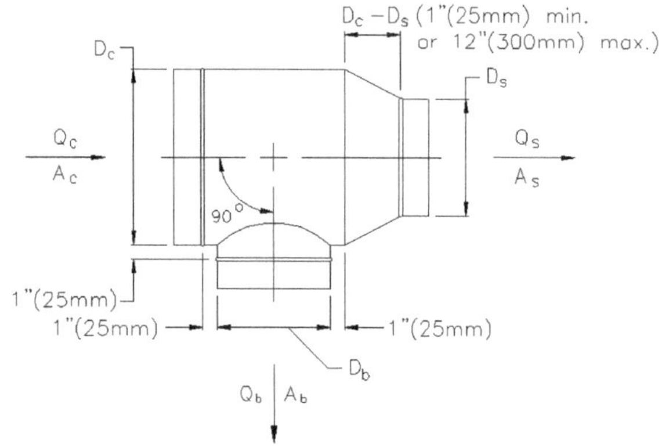

Diverging (fitting SD5-9)

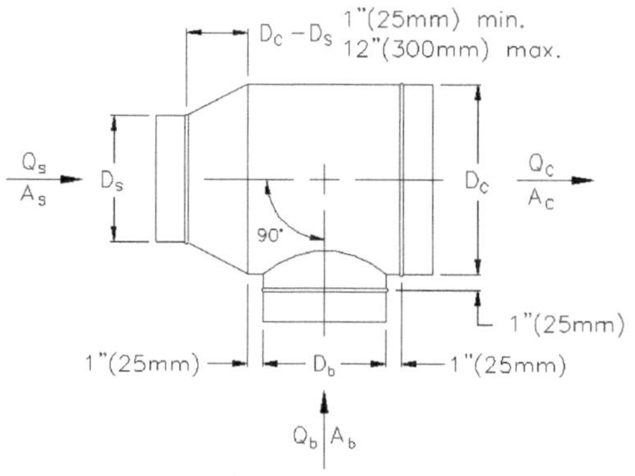

Converging (fitting ED5-3)

(a) Tee, Round, Straight Tap

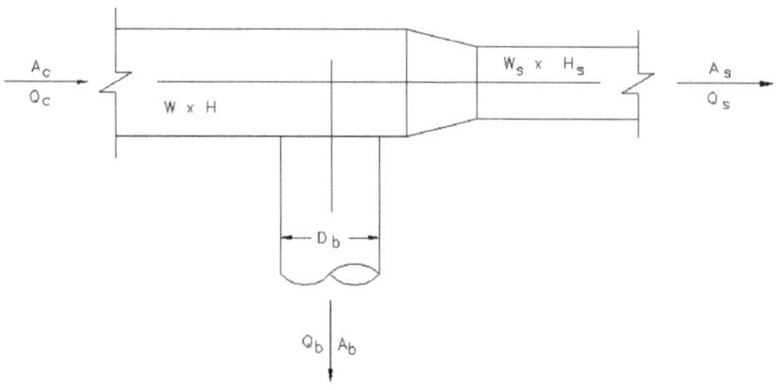

Diverging (fitting SR5-11)

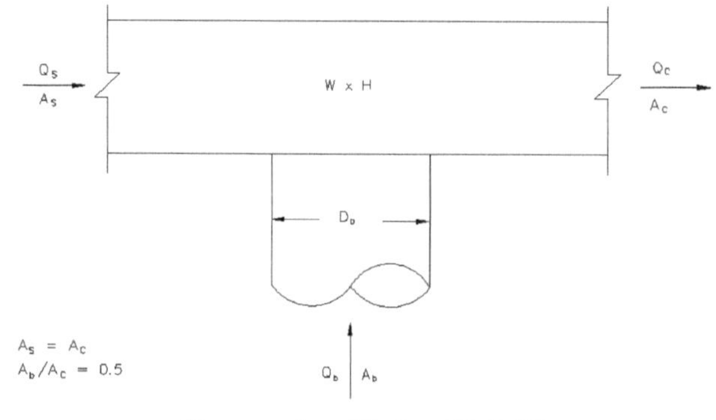

Converging (fitting ER5-2)

(b) Tee, Rectangular Main, Round Straight Tap

Figure 2-8 Tees for Low-Velocity Sections of Ductwork (2000 fpm Maximum)
(Reprinted from ASHRAE 2016)

Other recommended practices for flexible duct installation include the following:

- Install flexible duct hangers at least every 4 ft. Use flexible duct supports that are rigid and a minimum of 1.5 in. wide. Hangers should be of sufficient width to prevent any restriction of the internal diameter of the duct when the weight of the supporting section rests on the hanger.
- Install flexible duct so that sag does not exceed 1/2 inch per foot.
- Install flex duct so that bends exceed one duct diameter radius. Do not bend ducts across sharp corners such as pipes, wires, or conduits.
- Use only UL 181 flexible air ducts (UL 2017). Do not use flexible duct air connectors because they are not as durable and they are not tested for flame penetration, impact, or puncture. (Flexible *air ducts* and flexible *duct air connectors* are not interchangeable.)
- Use only metal straps. Nylon straps are not recommended until their problem of failure under high temperatures is solved.
- Do not install flexible duct upstream of VAV air terminal units because doing so adds turbulence, creating greater pressure drop (Stein et al. 2009, p. 3, item 23).
- Install flexible duct oversized to compensate for its additional resistance compared to that of round galvanized steel duct due to material roughness and routing (curves). Poorly installed flexible duct with many bends (some sharp with kinks), excessive compression, and sag is typically 15% to 20% compressed. Flexible duct pressure losses for various compression percentages can be determined using DFDB fitting CD11-2 (ASHRAE 2016) or by referring to the discussion of the pressure drop correction factor (PDCF) in Chapter 21, "Duct Design," of *ASHRAE Handbook—Fundamentals* (2017b).

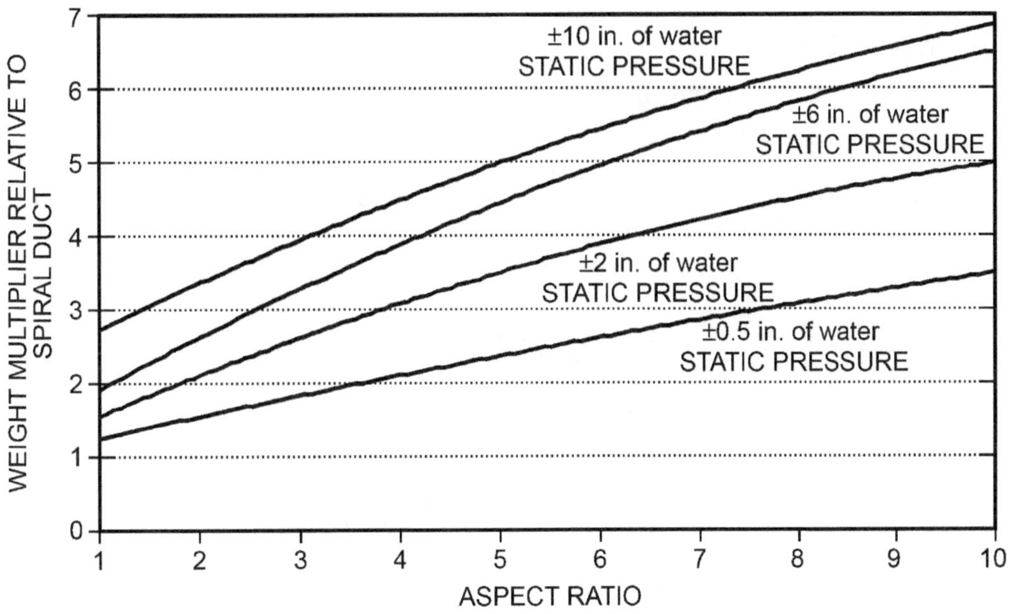

Notes: Based on SMACNA's Duct Construction Standards (2005). No scrap included. Includes reinforcement, and excludes hangers.

Figure 2-9 Relative Weights of Rectangular Duct to Round Spiral Duct
(Reprinted from ASHRAE 2017b, Figure 20)

BALANCING DAMPERS

Constant-Volume (CV) Systems

Dampers should be installed throughout CV systems, including systems designed using self-balancing methods. Install dampers at least three hydraulic diameters from any branches or changes in the direction of airflow. Design methods that are self-balancing, such as the equal friction method resized to balance pressure at each junction and the static regain method, produce systems that are fairly well balanced and that theoretically don't require balancing dampers. In reality, though, balancing dampers may be needed if close-coupled fittings are used, if fittings for which no data are available are used, or because of limitations in accuracy of the available fitting data (loss coefficients) (ASHRAE 2017b).

Variable-Air-Volume (VAV) Systems

Balancing dampers are not needed for VAV systems because the control dampers in VAV air terminal units balance the paths to provide the air quantity demand of each zone. VAV systems should have static pressure reset to prevent any air terminal unit control damper shutting too far and causing higher pressure losses that generate noise from the damper. Control algorithms can be used with static pressure data from the VAV air terminal units to adjust damper settings so that the VAV system stays at the minimum set point required to keep at least one control damper fully open at any given time (ASHRAE 2017b).

Riser takeoffs serving numerous floors from a single air handler may have large static pressure differences, so at each floor takeoff manual dampers should be provided. This will enable testing, adjusting, and balancing (TAB) contractors to field-adjust them after installation. The takeoff dampers may also be controlled dynamically to adjust the downstream static pressure of the VAV air terminal units. This will also drive the air handler to the lowest possible static pressure set point (ASHRAE 2013). Using reiterative computer static regain duct design software can sufficiently resize and balance a complex system to eliminate all balancing dampers under normal flow conditions.

Damper-Generated Noise

Consult Chapter 8 to avoid airflow-generated noise caused by dampers.

Extractors and Splitter Dampers

Extractors and splitter dampers (Figure 2-10) should never be used. They create noise and cause an increase in the pressure drop of the main section.

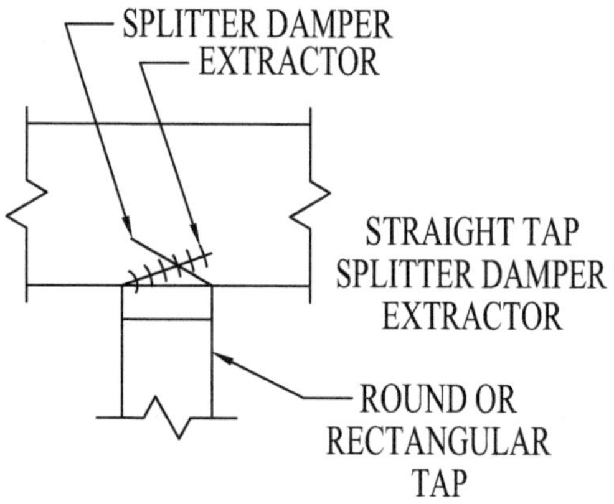

Figure 2-10 Straight Tap with Extractor or Splitter Tap

REFERENCES

Abushakra B., I.S. Walker, and M.H. Sherman. 2004. Compression effects on pressure loss in flexible HVAC ducts. *HVAC&R Research* 10(3):275–89.

ADC. 2010. *Flexible duct performance & installation standards*, 5th ed. Schaumburg, Illinois: Air Diffusion Council.

ASHRAE. 2013. Chapter 21, Duct design. In *ASHRAE handbook—Fundamentals*. Peachtree Corners, GA: ASHRAE.

ASHRAE. 2016. ASHRAE Duct Fitting Database, ver. 6.00.05 (online). Peachtree Corners, GA: ASHRAE. www.ashrae.org/technical-resources/bookstore/duct-fitting-database.

ASHRAE. 2017a. ANSI/ASHRAE Standard 120, *Method of testing to determine flow resistance of HVAC ducts and fittings*. Peachtree Corners, GA: ASHRAE.

ASHRAE. 2017b. Chapter 21, Duct design. In *ASHRAE handbook—Fundamentals*. Peachtree Corners, GA: ASHRAE.

ASHRAE. 2019. Chapter 49, Noise and vibration control. In *ASHRAE handbook—HVAC applications*. Peachtree Corners, GA: ASHRAE.

Culp, C. 2011. *HVAC flexible duct pressure loss measurements*. ASHRAE RP-1333 final report. Peachtree Corners, GA: ASHRAE.

Schaffer, M.E. 2005. *A practical guide to noise and vibration control for HVAC systems*, 2nd ed. (I-P). Peachtree Corners, GA: ASHRAE.

SMACNA. 2005. *HVAC duct construction standards, metal and flexible*, 3rd ed. Chantilly, VA: Sheet Metal and Air Conditioning Contractors' National Association.

Stein, J., A. Zhon, H. Cheng, M. Hydeman, S. Taylor, T. Hong, and J. Arent. 2009. *Advanced variable air volume (VAV) system design guide*, 2nd ed. Sacramento, CA: California Energy Commission.

UL. 2017. UL 181, *Standard for factory-made air ducts and air connectors*, 11th ed. Northbrook, IL: UL LLC.

3

Equal Friction Method

OVERVIEW

The equal friction method for sizing duct systems can be used for both supply and return/exhaust air systems for typical HVAC applications. Refer to Chapter 5 for local exhaust system design at constant velocity for particulate or fume collection systems.

As explained in Chapter 1, the total pressure loss in a duct system is comprised of the sum of the straight duct friction loss and the dynamic (fitting) loss of the individual duct sections. That concept is exemplified by Equation 3-1:

$$\Delta p_t = p_v\left[\left(\frac{12fL}{D_h}\right) + \left(\sum_{i=1}^{N} C_i\right)\right] \tag{3-1}$$

Though friction charts (which are approximate) and fundamental equations will provide friction loss data, the web-based ASHRAE Duct Fitting Database (DFDB; ASHRAE 2016) is the best source for determining the friction losses of straight ducts as well as the dynamic (fitting) losses. With the DFDB, round, flat oval, and rectangular straight duct can be selected, as can the absolute roughness and the air properties (air temperature, elevation, relative humidity, and viscosity). Refer to Chapter 21, "Duct Design," of *ASHRAE Handbook—Fundamentals* (2017), Chapter 1 of this design guide, or the DFDB (ASHRAE 2016) for a comprehensive list of absolute roughness factors. Using the DFDB is more accurate than just using a friction loss chart or a ductulator, because you can account for changes in the absolute duct roughness (for straight duct), actual air properties, and the velocity head. The DFDB automatically calculates the respective equivalent round diameter of rectangular and flat oval ducts to determine the friction loss. The DFDB can also calculate the fan system effect if there are fan inlet or outlet disruptions. If there are system effects caused by poor inlet and outlet fan configurations, the fan's installed performance will not match the fan at rated conditions. The fan system effect is accounted for in the calculations in the DFDB by including additional resistance categorized as "Fan & System Interactions" (Supply SR7-# and Exhaust/Return ED7-# series). More information on fan system effect can be found Chapter 6.

DUCT DESIGN BY THE EQUAL FRICTION METHOD

The equal friction methodology uses a constant (equal) friction rate. The maximum friction rate chosen should correspond to the maximum air velocity at which undesirable aerodynamic noise will not be generated; Table 3-1 is a good starting place for the initial design. Then, each sec-

Table 3-1 Maximum Recommended Duct Airflow Velocities to Achieve Specified Acoustic Design Criteria
(Reproduced from ASHRAE 2019, Table 8)

Duct Location	RC or NC Rating in Adjacent Occupancy	Maximum Airflow Velocity, fpm	
		Rectangular Duct	Round Duct
In shaft or above drywall ceiling	45	3500	5000
	35	2500	3500
	25 or less	1700	2500
Above suspended acoustic ceiling	45	2500	4500
	35	1750	3000
	25 or less	1200	2000
Within occupied space	45	2000	3900
	35	1450	2600
	25 or less	950	1700

Notes:
1. Branch ducts should have airflow velocities of about 80% of values listed.
2. Velocities in final runouts to outlets should be 50% of values or less.
3. Elbows and other fittings can increase airflow noise substantially, depending on type. Thus, duct airflow velocities should be reduced accordingly.

tion of ductwork is sized for the same friction rate. The Equal Friction Duct Design spreadsheet, available with this book at www.ashrae.org/DuctSyst, is based on calculating the results of Equation 3-1 for each section in the entire ductwork system (Chapter 1 includes a discussion of how to use this equation, which is provided in a different form as Equation 1-1). Consult Chapter 8 to determine how to calculate an acoustical analysis and for an explanation and definition of room criterion (RC). Limiting the maximum duct velocity to a recommended tabulated value per this spreadsheet does not replace conducting a complete ductwork system acoustical analysis for verifying that undesirable noise in the occupied space is avoided due to the fan selection, fittings, and dampers. The total pressure loss can be lowered by using lower friction rates, but this will result in larger duct sizes, which yields more expensive duct but lower fan energy costs.

When laying out a ductwork system it is important to avoid multiple fittings installed in close proximity to each other; both undesirable aerodynamic sound and unexpected turbulence will result in increased pressure drops. Best practice is to place fittings no closer than five hydraulic duct diameters apart (equations for hydraulic duct diameter are given in Chapter 1).

Determining the Friction Rate

To determine the friction rate in the equal friction design process, use the following steps:

1. Determine the maximum air velocity at the desired acoustic design criteria using Table 3-1.
2. Determine the air properties using Chapter 1, any of the many fluids mechanics and heat transfer textbooks that have extensive tabulations of air thermal properties, or using the "show air properties" radio button in the output section of the DFDB (ASHRAE 2016).

CD11-3 Straight Duct, Round, Maximum Velocity (Haaland 1983)
[Knowing Flow Rate (Q) and Target Velocity (V), determine Diameter (D)]

INPUT

Flow Rate (Q)	cfm	11400
Target Velocity (VT)	fpm	3500
Absolute Roughness (ei)	ft	0.00040
Density (RHO)	lbm/ft^3	0.061

Calculate

OUTPUT

Diameter (D)	in.	25.0
Velocity (V)	fpm	3,344
Velocity Pressure (Pv)	in. wg	0.57
Reynolds Number (Re)		584,044
Fricton Factor (f)		0.0150
Friction Rate (Po)	in. wg per 100 ft	0.41

Figure 3-1 Determining the Friction Rate Using the DFDB
(Reprinted from ASHRAE 2016)

3. Determine the duct absolute roughness (refer to Chapter 1 of this design guide, the Duct Design chapter of *ASHRAE Handbook—Fundamentals* [2017], or the DFDB [ASHRAE 2016]).
4. If you have access to the DFDB, select fitting CD11-3 and enter the flow rate and the maximum velocity from Table 3-1 (do not forget to select the appropriate air properties in My Account). For example, for round ductwork in a shaft with a desired noise criteria (NC) level of 35, the maximum recommended velocity is 3500 fpm. As shown in Figure 3-1, the duct that is included in Example 3-1 is designed for 11,400 cfm at an elevation of 5430 ft with an absolute roughness of 0.00040 ft. This results in a friction rate of 0.41 in. of water per 100 ft.
5. If you do not have access to the DFDB, the friction rate can be estimated using a friction loss chart at the desired velocity and airflow rate. In Figure 1-5 in Chapter 1, find the flow rate on the x axis. Then draw a vertical line to the velocity. Read the diameter from the diagonal line that crosses with the vertical line and read the friction rate from the y axis.

DUCT DESIGN SPREADSHEETS FOR USE WITH THE DFDB

The Microsoft® Excel® spreadsheets Equal Friction Duct Design and Equal Friction Duct Design Example, available with this book at www.ashrae.org/DuctSyst, are intended for use with the web-based DFDB when designing with the equal friction method. Example 3-1 shows how to use the spreadsheets to record pressure losses from the DFDB, calculate total path pressure losses, and determine balancing requirements. The pressure losses, air velocities, velocity pressures, and loss coefficients are taken directly from the DFDB (ASHRAE 2016). They are not calculated in the Equal Friction Duct Design Example spreadsheet; however, their summations are included.

Example 3-1.

Size the system shown in the figure below with the equal friction method using the DFDB. The design air temperature is 69°F and the building is located in Denver, Colorado, at 5430 ft elevation. Barometric pressure is 12.032 psia, relative humidity is 0% rh, density (ρ) is 0.061 lb_m/ft^3, there is zero duct air leakage, and the viscosity (μ) is 7.325×10^{-4} $lb_m/(ft \cdot min)$. Ducts are round spiral galvanized steel with an absolute roughness (ε) of 0.0004 ft. Assume the following:

- Distribution ductwork downstream of the VAV terminal unit is assumed to have a pressure loss of 0.05 in. of water.
- The flexible duct of section 20 in the figure below is 36.5 in. long at its installed length and 38 in. fully extended (4% compression).
- No dampers are required upstream of the VAV terminal units and the duct is sized for maximum flow conditions. The VAV terminal units will handle the change in airflow requirements.
- The first two sections are located in a shaft with an RC requirement of 35 maximum.
- The square-to-round duct off the fan is $H_1 = 27$ in., $W_1 = 20$ in., and length is 24 in.
- Sections 9, 10 and 17, 18 are bullhead tees with turning vanes.
- Sections 13 and 14 are a symmetrical wye with 45° elbows.
- Section 7 is a combination of a 45° wye with a 45° elbow that creates a 90° branch to the main.
- Section 12 includes a 45° wye.
- All other branches are 45° entry.
- The terminals are VAV terminal units and have loss coefficients according to the table below, derived from manufacturers' data for a basic assembly with two-row coils. The airflow has no allowance for air leakage.

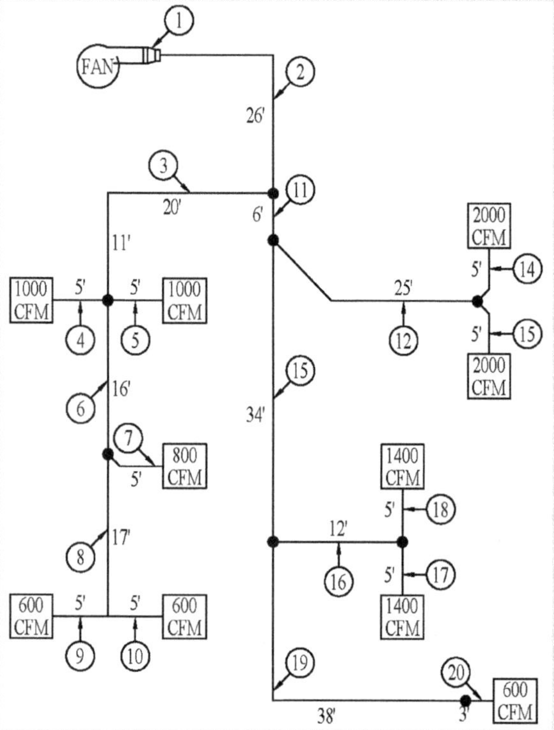

Duct Layout and Terminal Box Schedule

VAV Terminal Unit Resistance

Section	Box Size, in.	Airflow, cfm	Loss Coefficient C
4 and 5	10	1000	2.58
7	9	800	2.31
9 and 10	8	600	2.49
13 and 14	14	2000	2.56
17 and 18	12	1400	2.65
19	8	600	2.49

The total sum of airflow that must be supplied by the fan is the sum of the airflows required for the VAV terminal units plus the air leakage airflow. There is no diversity used in this example so the diversity factor is 1 (no adjustments to the airflow in any section are necessary). Using DFDB fitting CD11-3 at 3500 fpm maximum velocity and 11,400 total cfm, the friction rate to use for sizing is 0.41 in. of water per 100 ft (refer to Figure 3-1).

Use DFDB fitting CD11-4 to determine the diameter; it accounts for duct roughness and air properties. A friction loss chart (or a ductulator) could be used if adjusted for the design air properties; these also incorporate the correct duct roughness factor.

In this example, section 1 is the fan transition and section 2 is determined from the DFDB to be 25 in. in diameter. Refer to Figure 3-1, using fitting CD11-3 with the design airflows, air properties, and duct roughness, to determine the constant friction rate of 0.41 in. of water per 100 ft that is used in fitting CD11-4 to determine the diameter.

Solution.

The DFDB was used to determine fitting loss coefficients and calculate air density, velocity, velocity pressure, and duct pressure loss. In every instance the duct sizes were obtained using the friction rate of 0.41 in. of water per 100 ft and fitting CD11-4. The air quantity for each section is shown in the Equal Friction Example Problem and Summary of Individual Section Pressure Drops tables below, along with the air properties and duct roughness.

The sequence of calculations is as follows:

1. Section 2 consists of a 26 ft long duct that must convey 11,400 cfm of air from the fan to a total of 10 downstream VAV terminal units and a 90° round elbow.
2. Based on the maximum recommended duct airflow velocity to achieve a desired acoustic NC 35, DFDB fitting CD11-4 is used to calculate the diameter of section 2 as 25 in. with a design friction rate of 0.41 in. of water per 100 ft.
3. DFDB fitting CD11-1 is then used to calculate a total pressure loss of 0.11 in. of water for the straight-duct portion of section 2.
4. The 90° elbow in section 2 introduces no change in cross section—it is considered to be a part of section 2 (refer to the Sectioning a Duct System for Design section of Chapter 2. DFDB fitting CD4-9 gives a loss coefficient of 0.13 for the 90° elbow.
5. Section 1 is a required transition from the fan outlet to section 2, and it must likewise convey 11,400 cfm. DFDB fitting SD4-2 is used to calculate a loss coefficient of 0.01 (referenced to upstream conditions) for that transition. DFDB fitting SD4-2 output shows the calculation of a velocity pressure of 0.57 in. of water in the round section downstream of the transition. This likewise equals the velocity pressure in section 2, since the cross section of the straight duct and elbow in section 2 is equal to the downstream cross section in the transition of section 1.
6. The sum of loss coefficients in sections 1 and 2 is 0.14. When multiplied by the velocity pressure in those sections, the total pressure loss of the transition and elbow is 0.08 in. of water. When this total pressure loss is added to the total pressure loss of the straight duct, the total pressure loss of sections 1 and 2 is 0.19 in. of water. Section 11 consists of the main section of a tee with a 45° entry and 6 ft of straight duct. Section 11 must convey 7400 cfm of air to five downstream VAV terminal units. Therefore, DFDB fitting CD11-4 is used in conjunction with the design friction rate of 0.41 in. of water per 100 ft to calculate a duct diameter of 21 in. for section 11.
7. DFDB fitting CD11-1 is used to calculate a velocity pressure of 0.48 in. of water and a total pressure loss of 0.05 in. of water in the straight duct portion of section 11.

8. Based on a common section diameter of 25 in. (the diameter of section 2) and a main diameter of 21 in., as well as a common section flow rate of 11,400 cfm (from section 2), DFDB fitting SD5-12 is used to calculate a main section loss coefficient of 0.14. When multiplied by the velocity pressure in the main tee section, the total pressure loss of main section of the tee is 0.05 in. of water. When this total pressure loss is added to the total pressure loss of the straight duct, the total pressure loss of section 11 is 0.12 in. of water Section 3 consists of the branch of a tee with a 45° entry, 31 ft of straight duct, and a 90° round elbow. Each portion of this section has the same cross section, so these sections are treated as a single section. Section 3 must convey 4000 cfm of air to five downstream VAV terminal units, so DFDB fitting CD11-4 is used in conjunction with the design friction rate of 0.41 in. of water per 100 ft to calculate a duct diameter of 17 in.
9. DFDB fitting CD11-1 is then used to calculate a velocity pressure of 0.33 in. of water and a total pressure loss of 0.12 in. of water in the straight duct portion of section 3.
10. The straight-duct diameter of section 3 is also equal to the branch diameter of the tee; the common tee diameter (i.e., of section 2) is 25 in. Therefore, DFDB fitting SD5-12 gives a branch loss coefficient of 0.75 for section 3.
11. Likewise, DFDB fitting CD4-9 gives a loss coefficient of 0.15 for the 90° elbow in section 3. The sum of loss coefficients in section 3 is 0.90. When multiplied by the velocity pressure in section 3, the total pressure loss is 0.30 in. of water.

The remaining calculations proceed in a similar manner. The figure below summarizes the calculations, and the Individual Path Pressure Total Pressure Drops figure shows the total pressure drop of each path; the Imbalance figure summarizes the imbalance between paths. The imbalance (excess pressure) between paths needs to be balanced by the VAV terminal units in the noncritical paths. Alternatively, the designer could resize sections to make the paths more balanced. All duct velocities (see column 7 in the figure below) should be checked against Table 3-1 to make sure they meet the required RC level.

Equal Friction Duct Design Example - 3500 fpm

Air Temperature, °F	69	Relative Humidity, %	0
Elevation, ft	5430	Air Density, lbm/ft³	0.061
Barometric Pressures, psia	12.032	Viscosity (μ), lbm/(ft-min)	0.00073245

Upstream Section	Section	Fitting	ASHRAE Fitting Code	Air Quantity, cfm	Duct Size, in.	Velocity, fpm	Duct Length, ft	Velocity Pressure, p_v, n. of water	Loss Coefficient C	Total Pressure Loss, in. of water
		Source		Source					Source	
		Drawings	DFDB	Drawing	DFDB	DFDB	Drawing	DFDB	DFDB	Σ
1	2	Duct	CD11-3/CD11-1	11400	25	3344	26			0.11
		Elbow, 90°	CD3-9						0.13	
		Transition: H1=27.0",W1=20.0",L=24" (Theta1=5°,Theta2=12°)	SD4-2						0.01	
		Sized at Maximum Velocity of 3500 fpm						0.57	0.14	0.08
Section Total										**0.19**
2	11	Duct	CD11-1	7400	21	3077	11			0.05
		Tee, 45° Entry, Main (Dc=25, Ds=21, Db=17)	SD5-12						0.14	
								0.48	0.14	0.07
Section Total										**0.12**
11	15	Duct	CD11-1	3400	16	2435	23			0.09
		Wye, 45° (Dc=21, Ds=16, Db=17)	SD5-1						0.14	
								0.30	0.14	0.04
Section Total										**0.13**
15	19	Duct	CD11-1	600	8	1719	38			0.18
		Tee,45° Entry, Main (Dc=16, Ds=8, Db=15)	SD5-12						0.23	
		Elbow 90°	CD3-9						0.23	
								0.15	0.46	0.07
Section Total										**0.25**

Equal Friction Duct Design Example

(Excerpt from the Equal Friction Duct Design Example spreadsheet available with this book online)

Summary of Individual Section Pressure Drops

Section	Δp_t, in. of water
1/2	0.19
3	0.42
4 and 5	0.67
6	0.09
7	0.57
8	0.08
9 and 10	0.52
11	0.12
12	0.33
13 and 14	0.61
15	0.13
16	0.15
17 and 18	0.61
19	0.25
20	0.46

PATH A/B:		Path C:		Path D/E:		Path F/G:		Path H/I:		Path J:	
Section	TP, in. of water	Section	TP, in. of water	Section	TP, in. of water	Section	TP, in. of water	Section	TP, in. of water	Section	TP, in. of water
1/2	0.19	1/2	0.19	1/2	0.19	1/2	0.19	1/2	0.19	1/2	0.19
3	0.42	3	0.42	3	0.42	11	0.12	11	0.12	11	0.12
4/5	0.67	6	0.09	6	0.09	12	0.33	15	0.13	15	0.13
Total	1.28	7	0.57	8	0.08	13/14	0.61	16	0.15	19	0.25
		Total	1.27	9/10	0.52	Total	1.26	17/18	0.61	20	0.46
				Total	1.30			Total	1.20	Total	1.15

Individual Path Pressure Total Pressure Drops

(Excerpt from the Equal Friction Duct Design Example spreadsheet available with this book online)

Path to Terminal Box	TP, in. of water	Excess Pressure, in. of water	% Deviation
(4/5)	1.28	0.02	1.3%
(7)	1.27	0.03	2.1%
(9/10)	1.30	0.00	0.0%
(13/14)	1.26	0.04	3.1%
(17/18)	1.20	0.09	7.1%
(20)	1.15	0.14	11.1%

Imbalance

(Excerpt from the Equal Friction Duct Design Example spreadsheet available with this book online)

As can be seen from the Imbalance table, the critical path is path D or E ending in sections 9 or 10, respectively. The excess pressure column shows how much pressure will need to be dampened by the other VAV terminal units at maximum design conditions for the entire system to be balanced.

Design Steps

The design steps for designing a system with the equal friction method using the DFDB are as follows:

1. Lay out a single-line drawing of the system and assign section numbers as outlined in the Sectioning a Duct System for Design section of Chapter 2.
2. Locate balancing dampers in compliance with the Balancing Dampers section of Chapter 2.
3. Determine the air leakage in each section of ductwork and add these to the design air quantity required that resulted from the load calculations and system diversity (if applicable). (Diversity is the ratio of the actual duct system flow rate to the total system capacity and is typically expressed as a percentage.)
4. Determine terminal total pressure requirements for constant-volume diffusers or variable-air volume (VAV) terminal units. If required, calculate the loss coefficients per the DFDB CD8-9 to CD8-14 Terminal Unit (Box) series (ASHRAE 2016).
5. When using the DFDB, make sure to assign the correct air properties for each design.
6. Determine the constant friction rate to size duct sections:
   - Use the noise criteria (Table 3-1) as shown in step 4 of the Determining the Friction Rate subsection earlier in this chapter using DFDB fitting CD11-3 or
   - Use the following guidelines to select a friction rate:
     - If prevailing energy cost is high or installation labor cost is low, use 0.08 to 0.15 in. of water per 100 ft.
     - If prevailing energy cost is low or installation labor cost is high, use 0.30 to 0.60 in. of water per 100 ft.
     - Perform an acoustical analysis (refer to Chapter 8) to determine whether the noise levels are too high. If they are, you may need to lower the friction rate to eliminate or minimize airflow-generated (aerodynamic) noise to meet room design noise levels. An option is to provide lined or double-wall ductwork and/or duct sound attenuators (silencers) to reduce the noise levels.
7. Size all main and branch ducts at the constant friction rate using DFDB fitting CD11-4 or a friction loss chart. Use the duct size increments readily available per the manufacturer. Usually smaller sizes (less than 20 in. diameter) of spiral duct are available in 1-in. increments and larger sizes are available in 2-in. increments. Check with local suppliers.
8. Select efficient (low pressure drop) fittings for branches and changes of direction.
9. Using the Equal Friction Duct Design spreadsheet (available with this book at www.ashrae.org/DuctSyst and shown in Figure 3-2) with the DFDB, calculate the total pressure loss for each section separately for the supply and return ductwork. For each main and branch of a junction, be sure to account for the straight-through and branch loss coefficients. If you do not have access to the DFDB, the calculations must be done manually but can still be entered where the spreadsheet calls for the DFDB values.

Equal Friction Example Problem (DDG) [I-P]

Air Temperature, °F		Relative Humidity, %								
Elevation, ft		Air Density, lbm/ft^3								
Barometric Pressures, psia		Viscosity (μ), lbm/(ft-min)								
Upstream Section	Section	Fitting	ASHRAE Fitting Code	Air Quantity (cfm)	Duct Size (in.)	Velocity (fpm)	Duct Length (ft)	Velocity Pressure, p_v (in. wg)	Loss Coefficient, C	Total Pressure Loss (in. wg)
		Source		Source					Source	
		Drawings	DFDB	Drawing	DFDB	DFDB	Drawing	DFDB	DFDB	Σ
									0.00	0.00
Section Total										0.00

Figure 3-2 Equal Friction Duct Design Spreadsheet for Use with the DFDB
(From the Equal Friction Duct Design spreadsheet available with this book online)

10. Tabulate the total pressure required for each path from the fan to each respective supply and return terminal.
11. Determine the maximum operating pressure, then calculate the excess total pressure at each terminal.
12. When balancing is necessary to get within the accuracy of the calculations (0.1 in. of water), consider using smaller diameters of ductwork in noncritical paths (paths with excess pressure).
13. A further refinement is to use less-efficient fittings in noncritical paths. For non-VAV designs, balancing dampers need to be placed to provide the final balancing during field test and balancing (TAB).
14. Perform an acoustical analysis of the system (Chapter 8). Add lined duct or duct sound attenuators (silencers) where necessary.
15. If the system needs to be modified with different sizes to meet acoustical requirements or for other reasons, repeat steps 6 through 8.

DUCT DESIGN WITH OTHER DUCT SHAPES

Chapter 1 discusses the use of round versus flat oval and rectangular ductwork. Round duct systems are the most efficient from weight/cost/pressure/energy loss standpoints. Flat oval duct is closer to round than is rectangular, so it has many of the benefits of round along with some of the deficiencies of rectangular. Whatever shape is finally used, an analysis is normally designed assuming round duct and then the duct is converted to the desired shape by using the equivalent diameter equations (Equations 1-7 and 1-8 in Chapter 1).

Example 3-2 illustrates the calculation steps necessary if the round duct downstream of the fan (section 2) in Example 3-1 is replaced with a rectangular duct. Example 3-3 illustrates the calculation steps needed if the round duct downstream of the fan (section 2) in Example 3-1 is replaced with a flat oval duct.

Example 3-2.
Size the system depicted in Example 3-1 with the round duct downstream of the fan (section 2) replaced with a rectangular duct. Assume section 2 has a height restriction such that H must be 19 in. or less.

Solution.
Example 3-1 sized section 2 at 25 in. diameter. Equation 1-7 of Chapter 1 could be solved iteratively for W, given H = 18 in. and the equivalent round diameter D_e = 25 in.:

$$D_e = \frac{1.30AR^{0.625}}{P^{0.250}} = \frac{1.30(WH)^{0.625}}{(W+H)^{0.250}}$$

Alternatively, the "Rectangle Equiv Round" tab in the Equivalent Round Tables spreadsheet available with this book at www.ashrae.org/DuctSyst can be used. The figure below shows part of that spreadsheet for rectangular duct. Using interpolation, the major axis is 30 in.

Major[2] Axis W, inch	Minor[1] Axis H, Inch																
	3	4	5	6	7	8	9	10	11	12	14	16	18	20	22	24	30
8	5	6	7	8	8												
10	6	7	8	8	9	10	10										
12	6	7	8	9	10	11	11	12	13								
14	7	8	9	10	11	11	12	13	14	14							
16	7	8	9	10	11	12	13	14	14	15	16						
18	7	9	10	11	12	13	14	15	15	16	17	19					
20	8	9	10	11	13	13	14	15	16	17	18	20	21				
22	8	9	11	12	13	14	15	16	17	18	19	20	22	23			
24	8	10	11	12	14	15	16	17	17	18	20	21	23	24	25		
28	9	10	12	13	14	16	17	18	19	20	21	23	24	26	27	28	
32	9	11	13	14	15	17	18	19	20	21	23	24	26	27	29	30	34
36	10	12	13	15	16	17	19	20	21	22	24	26	27	29	31	32	36
40	10	12	14	15	17	18	19	21	22	23	25	27	29	30	32	34	38
44	11	13	14	16	17	19	20	22	23	24	26	28	30	32	34	35	40

Excerpt from the "Rect Equiv Round" tab of the Equivalent Round Tables Spreadsheet

Section 2 pressure losses can then be determined. A rectangular-to-round transition must be added to the end of section 2 as it transitions to section 11. The following figure shows the data from the Equal Friction Duct Design Example spreadsheet available with this book online.

Air Temperature, °F		69	Relative Humidity, %	0						
Elevation, ft		5430	Air Density, lbm/ft^3	0.061						
Barometric Pressures, psia		12.032	Viscosity (μ), lbm/(ft-min)	0.00073245						
Upstream Section	Section	Fitting	ASHRAE Fitting Code	Air Quantity, cfm	Duct Size, in.	Velocity, fpm	Duct Length, ft	Velocity Pressure, p_v, n. of water	Loss Coefficient C	Total Pressure Loss, in. of water
		Source		Source					Source	
		Drawings	DFDB	Drawing	DFDB	DFDB	Drawing	DFDB	DFDB	Σ
1	2	Duct, H=18", W=30"	CR11-1	11400	H=18, W=30	3040	26			0.1
		Elbow, Mitered w 2 1/8-in Van Spacing, 90°	CR3-15						0.25	
		Fan Transition: H1=27.0",W1=20.0",Ho=18.0" Wo=30", L=24" θ1 = 21 deg. θ2 =24 deg.	SR4-2						0.00	
		Rectangular to Round Transition Section 11: H=18", W=30", Do=21, L=24"	SD4-2						0.1	
		Sized for De = 25, H=18"						0.47	0.35	0.16
Section Total										0.26

Revision of Section 2 for a Height Restriction of H = 18 in. Using Rectangular Duct
(Excerpt from the Equal Friction Duct Design Example spreadsheet available with this book online)

This increased the pressure loss for section 2. However, the acoustical criteria are no longer met. Refer to Table 3-1 for the maximum allowable velocity in a shaft to meet the RC35 requirement: 2500 fpm. The width W needs to increase to 37 in.:

Air Temperature, °F		69	Relative Humidity, %	0						
Elevation, ft		5430	Air Density, lbm/ft^3	0.061						
Barometric Pressures, psia		12.032	Viscosity (μ), lbm/(ft-min)	0.00073245						
Upstream Section	Section	Fitting	ASHRAE Fitting Code	Air Quantity, cfm	Duct Size, in.	Velocity, fpm	Duct Length, ft	Velocity Pressure, p_v, n. of water	Loss Coefficient C	Total Pressure Loss, in. of water
		Source		Source					Source	
		Drawings	DFDB	Drawing	DFDB	DFDB	Drawing	DFDB	DFDB	Σ
1	2	Duct, H=18", W=37"	CR11-1	11400	H=18, W=37	2465	26			0.06
		Elbow, Mitered w 2 1/8-in Van Spacing, 90°	CR3-15						0.25	
		Fan Transition: H1=27.0",W1=20.0",Ho=18.0" Wo=37", L=24" θ1 = 21 deg. θ2 =24 deg.	SR4-2						0.02	
		Rectangular to Round Transition Section 11: H=18", W=30", Do=21, L=24"	SD4-2						0.19	
		Sized for De = 25, H=18"						0.31	0.46	0.14
Section Total										0.20

Revision of Section 2 for a Height Restriction of H = 18 in. Using Rectangular Duct so the Section's Velocity Does Not Exceed 2500 fpm

The rest of the system did not change, and because section 2 is common to all paths, the difference between the original design for section 2 and this one can be added to the original total pressure drop. Note that with the design at the same velocity, the rectangular first section pressure loss was 0.26 in. of water compared to the original for the round section 2 of 0.19 in. of water. When the velocity was reduced to a maximum of 2500 fpm, the difference just added 0.01 in. of water. Therefore, 0.01 in. of water can be added to the round design to get the total pressure drop for the design with the rectangular second section. The total pressure drop or critical path for Example 3-1 is 1.62 in. of water. Changing the second section to 18 in. × 37 in. only increased the total pressure requirement to 1.63 in water. The imbalance did not change since section 2 is common to all paths.

Example 3-3.

Size the system depicted in Example 3-1 assuming section 2 has a height restriction such that the minor *a* dimension of the oval duct must be 18 in. or less in height. Equation 1-8 of Chapter 1 can be solved iteratively for *A* for the given $D_e = 25$ in.:

$$D_e = \frac{1.55AR^{0.625}}{P^{0.250}} = \frac{1.55\left[\frac{\pi}{4}a + a(A-a)\right]^{0.625}}{[\pi a + 2(A-a)]^{0.250}}$$

Solution.

Example 3-1 sized section 2 at 25 in. diameter. The above equation could be solved for the major dimension *A* iteratively with $a = 18$ in. and $D_e = 25$ in., or the "Flat Oval Equiv Round" tab of the Equivalent Round Tables spreadsheet, available with this book at www.ashrae.org/DuctSyst, in could be used. Using this spreadsheet, the flat oval duct major dimension is calculated as 32 in. (see figure below).

Major[2] Axis W, inch	Minor[1] Axis H, inch: 3	4	5	6	7	8	9	10	11	12	14	16	18	20	22	24	30
8	5	6	7	8	8												
10	6	7	8	8	9	10	10										
12	6	7	8	9	10	11	11	12	13								
14	7	8	9	10	11	11	12	13	14	14							
16	7	8	9	10	11	12	13	14	14	15	16						
18	7	9	10	11	12	13	14	15	15	16	17	19					
20	8	9	10	11	13	13	14	15	16	17	18	20	21				
22	8	9	11	12	13	14	15	16	17	18	19	20	22	23			
24	8	10	11	12	14	15	16	17	17	18	20	21	23	24	25		
28	9	10	12	13	14	16	17	18	19	20	21	23	24	26	27	28	
32	9	11	13	14	15	17	18	19	20	21	23	24	26	27	29	30	34
36	10	12	13	15	16	17	19	20	21	22	24	26	27	29	31	32	36
40	10	12	14	15	17	18	19	21	22	23	25	27	29	30	32	34	38

Excerpt from the "Flat Oval Equiv Round" tab of the Equivalent Round Tables Spreadsheet

Section 2 pressure losses can then be determined. A flat oval to round transition was added to the end of section 2 as it transitions to section 11. The following figure shows the data from the Equal Friction Duct Design Example spreadsheet available with this book online.

Air Temperature, °C	69	Relative Humidity, %	0
Elevation, m	5430	Air Density, lbm/ft³	0.061
Barometric Pressures, psia	12.032	Viscosity (μ), lbm/(ft-min)	0.00073245

Upstream Section	Section	Fitting	ASHRAE Fitting Code	Air Quantity, cfm	Duct Size, in.	Velocity, fpm	Duct Length, ft	Velocity Pressure, p_v, n. of water	Loss Coefficient C	Total Pressure Loss, in. of water
		Source		Source					Source	
		Drawings	DFDB	Drawing	DFDB	DFDB	Drawing	DFDB	DFDB	Σ
1	2	Flat Oval Duct	CF11-1	11400	a=18, A=32	3241	26			0.1
		Elbow, 90°	CF3-1						0.47	
		Transition: H1=27.0",W1=20.0",L=24" (Theta1=28°,Theta2=6°), Used De as round size since this fitting does not exist in the DFDB	SD4-2						0.02	
		Transition Flat Oval to Round: A=27.0",a=20.0",L=24", Do=21	SF4-1						0.05	
		Sized at Maximum Velocity of 3500 fpm						0.54	0.52	0.28
Section Total										0.38

Revision of Section 2 for a Height Restriction of *a* = 18 in. using Flat Oval Duct

(Excerpt from the Equal Friction Duct Design Example spreadsheet available with this book online)

This increased the pressure loss for section 2 from 0.19 to 0.38 in. of water. The rest of the system did not change, since section 2 is common to all paths. The difference between the original design for section 2 and this one can be added to the original total pressure drop. So, 0.19 in. of water can be added to the original round design to get the total pressure drop for the design with the flat oval section 2. That is, the total pressure drop or critical path for Example 3-1 is 1.67 in. of water. Changing the second section to 18 in. × 32 in. increased the total pressure requirement to 1.80 in water. The imbalance did not change, since section 2 is common to all paths. Section 2 pressure losses can then be determined. Note a flat oval to round transition must be added to the end of section 2 as it transitions to section 11.

NOMENCLATURE

A = major axis of flat oval duct, in.
a = minor axis of flat oval duct, in.
C = fitting loss coefficient, dimensionless
D = diameter, in.
D_e = circular equivalent of rectangular or flat oval duct for equal length, fluid resistance, and airflow, in.
D_h = hydraulic diameter, in.
f = friction factor, dimensionless
H = length of adjacent side of rectangular duct, in.
L = duct length, ft
N = number of fittings in a duct section
P = duct perimeter used for equivalent round diameter, in.
p_v = velocity pressure, in. of water
W = length of one side of rectangular duct, in.

Symbols

Δp_t = total pressure loss, in. of water
ε = absolute roughness
μ = viscosity, $lb_m/(ft \cdot min)$
ρ = density, lb_m/ft^3
θ = angle

Subscripts

i = the ith fitting in a section

REFERENCES

ASHRAE. 2016. ASHRAE Duct Fitting Database, ver. 6.00.05 (online). Peachtree Corners, GA: ASHRAE. www.ashrae.org/technical-resources/bookstore/duct-fitting-database.

ASHRAE. 2017. Chapter 21, Duct design. In *ASHRAE handbook—Fundamentals*. Peachtree Corners, GA: ASHRAE.

ASHRAE. 2019. Chapter 49, Noise and vibration control. In *ASHRAE handbook—HVAC applications*. Peachtree Corners, GA: ASHRAE.

Schaffer, M.E. 2005. *A practical guide to noise and vibration control for HVAC systems*, 2nd ed. (I-P). Peachtree Corners, GA: ASHRAE.

4

Static Regain Method

OVERVIEW

The static regain duct design method is applicable only to supply air systems. This methodology uses the conservation of momentum principle which results in an increase in static pressure when the velocity is reduced in an airstream (see Figure 4-1). The method sizes the main duct and branches between takeoffs so that the recovery in static pressure between takeoffs due to the reduction in velocity is approximately equal to the total pressure loss due to friction and fittings in this duct section according to Equation 4-1 or 4-2. For Equation 4-1, a negative static regain indicates that the reduction in velocity pressure is greater than the total pressure loss.

Figure 4-1 illustrates the static regain design process for one section of a duct system. Sections 1 and 2 ducts are sized so that the change in velocity pressure ($p_{v1} - p_{v2}$) conforms to Equation 4-2:

$$p_{s2} - p_{s1} = \left[\rho\left(\frac{V_1}{1097}\right)^2 - \rho\left(\frac{V_2}{1097}\right)^2\right] - \Delta p_{t,\,1-2} \cong 0 \tag{4-1}$$

$$p_{s2} - p_{s1} = [p_{v1} - p_{v2}] - \Delta p_{t,\,1-2} \cong 0 \tag{4-2}$$

DUCT DESIGN BY THE STATIC REGAIN METHOD

To avoid generating undesirable aerodynamic noise in a duct system, use Table 3-1 in Chapter 3 as a guide for the design velocity of the initial section (the first supply air section after

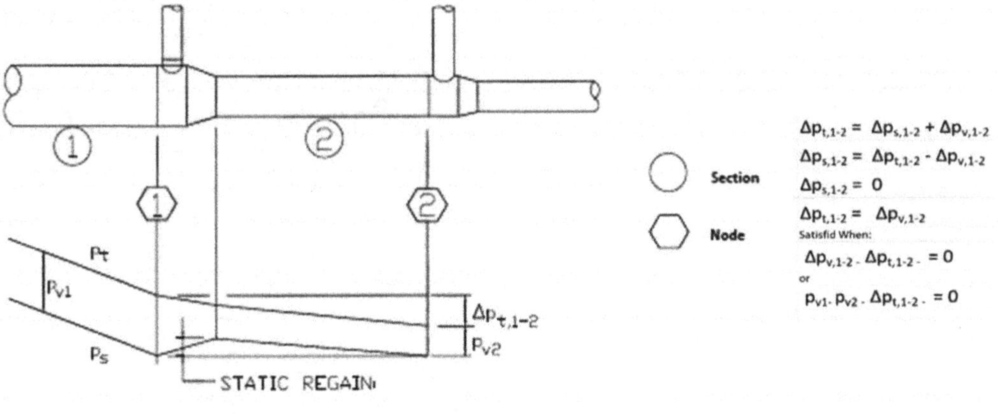

Figure 4-1 Static Regain Design Process

the fan). For system branches use the junction upstream velocity and design up to 1) the variable-air-volume (VAV) terminal boxes and/or the terminals with modulating diffusers or 2) the diffusers in a constant-volume (CV) system. Note that the use of efficient takeoffs (tees, laterals, crosses) can result in smaller duct sizes and lower operating costs. The terminal section pressure must be adequate to operate VAV terminal units with distribution ductwork, modulating diffusers, and CV diffusers. Design distribution ductwork downstream of VAV terminal units by the equal friction method (see Chapter 3).

Calculate the pressure loss for each path, which includes 1) the VAV terminal unit, distribution ductwork, and diffusers or 2) CV diffusers. The final design should be reasonably in balance, which means that the pressure loss of each path from the fan to the terminal should be reasonably close to that of the design leg(s). When VAV terminal units are used they should be able to handle any small imbalance that exits.

DUCT DESIGN SPREADSHEETS FOR USE WITH THE DFDB

The Microsoft® Excel® spreadsheets Static Regain Duct Design (shown in Figure 4-2) and Static Regain Duct Design Example, both available with this book at www.ashrae.org/DuctSyst, are intended for use with the web-based ASHRAE Duct Fitting Database (DFDB; ASHRAE 2016) when designing with the static regain method. Example 4-1 shows how to use the spreadsheets to record pressure losses from the DFDB, calculate total path pressure losses, and determine balancing requirements. The pressure losses, velocities, velocity pressures, and loss coefficients are taken directly from the DFDB; they are not calculated in the spreadsheets, although the summations are.

The design steps for designing a system with the static regain method using the DFDB are:

1. Lay out a single-line drawing of the system and assign section numbers as outlined in the Sectioning a Duct System for Design section of Chapter 2.
2. Locate balancing dampers in compliance with the Balancing Dampers section of Chapter 2.
3. Determine the air leakage in each section of ductwork and add these to the design air quantity required resulting from the load calculations and system diversity (if applicable).
4. Determine terminal total pressure requirements for CV diffusers or VAV terminal units. If applicable, calculate loss coefficients per the DFDB CD8-# Terminal Unit (Box) series.

Air Temperature, °F		Relative Humidity, %	
Elevation, ft		Air Density, lbm/ft^3	
Barometric Pressures, psia		Viscosity (μ), lbm/(ft-min)	

Upstream Section	Section	Fitting	ASHRAE Fitting Code	Air Quantity, cfm	Duct Size, in.	Velocity, fpm	Duct Length, ft	Velocity Pressure, p_v, in. of water	Loss Coefficient C	Total Pressure Loss, in. of water	Regain, in. of water $[p_{v1} - p_{v2}] - \Delta p_t$
		Source		Source				Source			Source
		Drawings	DFDB	Drawings	Iteration	DFDB	Drawings	DFDB	DFDB	Σ	Static Regain Calculation
									0.00	0.00	
Section Total										0.00	
									0	0.00	
Section Total										0.00	0.00

Figure 4-2 Static Regain Duct Design Spreadsheet for Use with the DFDB
(From the Static Regain Duct Design spreadsheet available with this book online)

5. Size the fan discharge duct (the first supply air section after the fan). Use the duct size increments readily available in the industry. Usually smaller sizes of spiral duct are available in 1-in. increments (less than 20 in. diameter) and larger sizes are available in 2-in. increments.
6. The maximum recommended initial duct velocity is given in Table 3-1 of Chapter 3. Limit duct velocity so the aerodynamically generated noise in ducts is no more than the room criterion (RC) values noted. (Consult Chapter 8 for an explanation and definition of RC.) An alternative for higher duct velocities is to provide lined ductwork and/or duct silencers to attenuate the increase in duct-generated noise.
7. Size the straight-through sections first using Equation 4-1 and the Static Regain Duct Design spreadsheet, available at www.ashrae.org/DuctSyst. This could take several iterations. Start with the same diameter as the upstream section. If the regain for the section is positive ($p_{s2} - p_{s1} > 0$), decrease the section size until the regain is negative, then use the previous size. The regain, or the change in static pressure from one junction to the next, should be zero or just positive. (Note that if you are manually sizing the ductwork and most of the air is going to the branch, you may want to skip sizes to reduce the number of calculations required.)
8. Size the branches using the same method, up to the VAV terminal units, if any. Use the junction *upstream* velocity as p_{v1}.
9. Size ductwork downstream of VAV terminal units using the equal friction method (see Chapter 3).
10. Tabulate the total pressure required for each path from the fan to each supply terminal and calculate the excess total pressure at each terminal. The design should be reasonably in balance. If not, adjust the appropriate branch by changing the duct size or fittings. Imbalance of 0.1 in. of water is acceptable (well within the accuracy of the fitting loss coefficients).
11. Perform an acoustical analysis of the system (consult Chapter 8). Provide lined duct or sound attenuators where necessary.

DUCT DESIGN WITH OTHER DUCT SHAPES

Static regain designs, like equal friction designs, are initially designed with round duct. Refer to Chapter 1 if other shapes like rectangular or flat oval are required.

Example 4-1.

Design the VAV system depicted in the figure below by the static regain method. The design air temperature is 69°F, and the building is located in Denver, Colorado, at 5430 ft elevation. Barometric pressure is 12.032 psia, relative humidity is 0%, density (ρ) is 0.061 lb_m/ft^3, and viscosity (μ) is 7.325 × 10^{-4} $lb_m/(ft{\cdot}min)$. Ducts are round spiral galvanized steel with an absolute roughness (ε) of 0.0004 ft. Several of these values are calculated in the DFDB. Assume the following:

- Distribution ductwork downstream of the VAV terminal unit is assumed to have a pressure loss of 0.05 in. of water.
- The flexible duct of section 20 is 36.5 in. at its installed length and 38 in. fully extended (4% compression).
- No dampers are installed upstream of the VAV terminal units (per recommended design practice), and the duct is sized for maximum flow conditions. The VAV terminal units will handle the change in airflow requirements.

- The first two sections are located in a shaft with an RC requirement of 35 maximum.
- The square-to-round duct off the fan is $H_1 = 27$ in., $W_1 = 20$ in., and length is 24 in.
- Section 9 and 10 and Section 17 and 18 are bullhead tees with turning vanes.
- Sections 13 and 14 are a symmetrical wye with 45° elbows.
- Section 7 is a combination of a 45° wye with a 45° elbow that creates a 90° branch to the main.
- Section 12 is a wye.
- All other branches are 45° entry.
- The terminals are VAV terminal units and have loss coefficients derived from manufacturers' data for a basic assembly with two-row coils (see the table below).

The airflow has no allowance for air leakage.

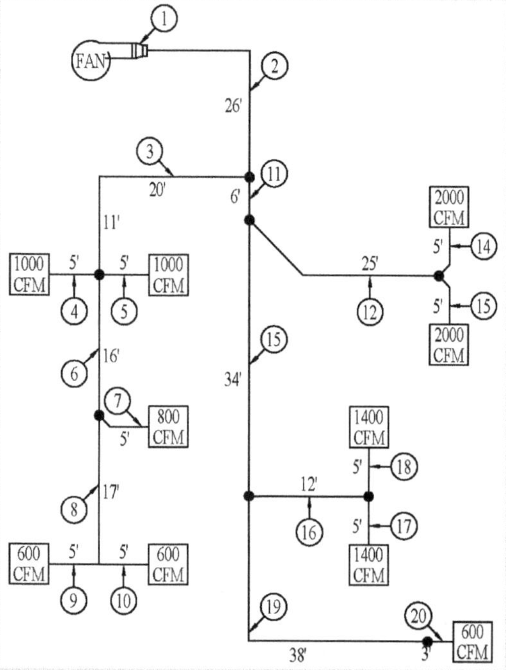

Duct Layout and Terminal Box Schedule

VAV Terminal Unit Resistance

Section	Box Size, in.	Airflow, cfm	Loss Coefficient C
4 and 5	10	1000	2.58
7	9	800	2.31
9 and 10	8	600	2.49
13 and 14	14	2000	2.56
17 and 18	12	1400	2.65
20	8	600	2.49

The total sum of airflow that must be supplied by the fan is the sum of the airflows required for the VAV terminal units plus the air leakage airflow. There is no diversity used in this example so the diversity factor is 1 (no adjustment to the airflow in any section is necessary). The fan transition is fixed. Section 2 is sized at the maximum velocity of 3500 fpm from Table 3-1 of Chapter 3 per assumption (d); 1-in. duct increments are used.

Solution.

The figure below shows the iterations and results of a static regain analysis for the example system. Straight-through main sections are treated first (sections 11, 15, 19, and 20), followed by the branches off the main sections (sections 3 and 12), followed by straight-throughs (sections 6 and 8), then the remainder of the branches (sections 4, 5, 9, 10, 13, 14, 17, and 18). An asterisk by a section number in the figure below indicates the selected sizes. The DFDB (ASHRAE 2016) was used to determine fitting loss coefficients, air density, velocity, velocity pressure, and duct pressure loss.

The following figure also summarizes the excess pressure for each path. Most of the excess pressures are less than or about equal to 0.10 in. of water.

Air Temperature, °F	69	Relative Humidity, %	0
Elevation, ft	5430	Air Density, lbm/ft³	0.061
Barometric Pressures, psia	12.032	Viscosity (μ), lbm/(ft-min)	0.00073245

Upstream Section	Section	Fitting	ASHRAE Fitting Code	Air Quantity, cfm	Duct Size, in.	Velocity, fpm	Duct Length, ft	Velocity Pressure, p_v, in. of water	Loss Coefficient C	Total Pressure Loss, in. of water	Regain, in. of water $[p_{v1} - p_{v2}] - \Delta p_t$
		Source Source		Source				Source			Source
		Drawings	DFDB	Drawings	Iteration	DFDB	Drawings	DFDB	DFDB	Σ	Static Regain Calculation
1	2*	Duct	CD11-1	11400	25	3334	26			0.11	
		Elbow	CD3-9						0.13		
		Transition: H1= 20", W1= 27", L=24" (Theta1=17°, Theta2=0°)	SD4-2						0.01		
		Sized at Maximum Velocity of 3500 fpm						0.57	0.14	0.08	
Section Total										0.19	
2	11a	Duct	CD11-1	24	25	2171	11			0.02	
		Tee, 45° Entry, Main: Dc=25, Ds=25, Db=25	SD5-1						0.15		
								0.24	0.15	0.04	(0.57 - 0.24) - 0.06= 0.27
Section Total										0.06	0.27
2	11b	Duct	CD11-1	7400	24	2355	11			0.02	
		Tee, 45° Entry, Main: Dc=25, Ds=24, Db=24	SD5-12						0.15		
								0.28	0.15	0.04	(0.57-0.28)-0.06 =0.23
Section Total										0.06	0.23

Static Regain Duct Design Example

(Excerpt from the Static Regain Duct Design Example spreadsheet available with this book online)

The following figure shows the results of the iterative calculations.

Section	ΔP_t, in. of water
1	0.0
2	0.19
3	0.37
4 & 5	0.61
6	0.04
7	0.53
8	0.03
9 &10	0.48
11	0.10
12	0.26
13 & 14	0.54
15	0.10
16	0.11
17 & 18	0.56
19	0.15
20	0.49

Summary of Individual Section Pressure Drops

(Excerpt from the Static Regain Duct Design Example spreadsheet available with this book online)

The following figure summarizes the total pressure losses of each path in the system. The critical path is path A or B ending in sections 4 or 5, respectively.

Path A/B:		Path C:		Path D/E:		Path F/G:		Path H/I:		Path J:	
Section	TP, in. of water	Section	TP, in. of water	Section	TP, in. of water	Section	TP, in. of water	Section	TP, in. of water	Section	TP, in. of water
1	0.0	1	0.0	1	0.0	1	0.0	1	0.0	1	0.0
2	0.19	2	0.19	2	0.19	2	0.19	2	0.19	2	0.19
3	0.37	3	0.37	3	0.37	11	0.10	11	0.10	11	0.10
4/5	0.61	6	0.04	6	0.04	12	0.26	15	0.10	15	0.10
Total	1.16	7	0.53	8	0.02	13/14	0.54	16	0.11	19	0.15
		Total	1.12	9/10	0.48	Total	1.08	17/18	0.56	20	0.49
				Total	1.09			Total	1.06	Total	1.04

Individual Path Pressure Total Pressure Drops

(Excerpt from the Static Regain Duct Design Example spreadsheet available with this book online)

As seen in the preceding table, the critical path (the path with the highest overall pressure loss) is path A or B ending in sections 4 or 5, respectively. The excess pressure column shows how much pressure will need to be dampened by the other VAV terminal units at maximum design conditions for the entire system to be balanced. Because the imbalance, or excess pressure, in the noncritical paths is very low, the VAV terminal units should be able keep the system in balance without additional dampening. (See the following figure for the imbalance.)

Path to Terminal Box	TP, in. of water	Excess Pressure, in. of water	% Deviation
(4/5)	1.16	0.00	0
(7)	1.12	0.04	4
(9/10)	1.09	0.07	6
(13/14)	1.08	0.08	7
(17/18)	1.06	0.11	9
(20)	1.04	0.13	11

Imbalance

(Excerpt from the Static Regain Duct Design Example spreadsheet available with this book online)

No redesign is necessary for most paths because their excess pressures are about 0.1 in. of water or less. Path J, however, has an excess pressure of 0.13 in. of water, so there could be some redesign for that path. By decreasing section 19 to 8 in. diameter, the excess pressure is reduced to 0.06 in. of water without increasing the total pressure. The above table shows the decrease in the imbalance from resizing section 19 to 8 in. diameter.

The following figure compares the total pressure losses of each path in the system and indicates that the static regain method has produced a more balanced system than the equal friction method.

Path to Terminal Box	TP, in. of water	Excess Pressure, in. of water	% Deviation
(4/5)	1.16	0.00	0
(7)	1.12	0.04	4
(9/10)	1.09	0.07	6
(13/14)	1.08	0.08	7
(17/18)	1.06	0.11	9
(20)	1.10	0.06	6

Imbalance with Section 19 Resized

(Excerpt from the Static Regain Duct Design Example spreadsheet available with this book online)

All duct velocities are acceptable because they are less than the maximum allowable velocity for RC = 35 (compare the calculated air velocity for each duct section in the first table of this example against column 4 of Table 3-1).

NOMENCLATURE

C = fitting loss coefficient, dimensionless
D = diameter, in.
H = length of adjacent side of rectangular duct, in.
L = duct length, ft
p_{s2} = static pressure upstream of junction 2, in. of water
p_{s1} = static pressure upstream of junction 1, in. of water
p_t = total pressure, in. of water
$\Delta p_{t,1-2}$ = total pressure loss from junction 1 to junction 2, in. of water
p_{v1} = velocity pressure at section 1, in. of water

p_{v2} = velocity pressure at section 2, in. of water
V_1 = velocity upstream at section 1, fpm
V_2 = velocity upstream at section 2, fpm
W = length of one side of rectangular duct, in.

Symbols

ε = absolute roughness, ft
μ = viscosity, $lb_m/(ft \cdot min)$
ρ = density, lb_m/ft^3

Subscripts

b = straight section
c = common section
s = straight section

REFERENCES

ASHRAE. 2016. ASHRAE Duct Fitting Database, ver. 6.00.05 (online). Peachtree Corners, GA: ASHRAE. www.ashrae.org/technical-resources/bookstore/duct-fitting-database.

ASHRAE. 2019. Chapter 49, Noise and vibration control. In *ASHRAE handbook—HVAC applications*. Peachtree Corners, GA: ASHRAE.

Schaffer, M.E. 2005. *A practical guide to noise and vibration control for HVAC systems*, 2nd ed. (I-P). Peachtree Corners, GA: ASHRAE.

5

Constant Velocity Method

OVERVIEW

The constant velocity duct design method is applicable only to local exhaust/ventilation systems, and this chapter covers using this method for local exhaust systems conveying fumes, dust, particulates, and contaminants from various processes.

A typical local exhaust ventilation systems consist of the following basic elements:

- Hood to capture pollutants, vapors, and/or excessive heat
- Ducts to transport polluted air to an air-cleaning device or vent the exhaust air from the building
- Air-cleaning device to remove captured pollutants from the airstream for recycling or disposal
- Air-moving device (e.g., fan or high-pressure air ejector) to provide motive power to overcome system resistance
- Exhaust stack to discharge system air to the atmosphere

These elements are covered in detail by the following chapters of the American Conference of Governmental Industrial Hygienists (ACGIH) publication *Industrial Ventilation: A Manual of Recommended Practice for Design* (2019), or *ASHRAE Handbook—HVAC Applications* (2019):

Hoods:	ACGIH chapters 6 and 13
Air-cleaning devices:	ACGIH chapter 8
Fans:	ACGIH chapter 7
Stack design:	ACGIH chapter 5, Section 5.12, and/or ASHRAE chapter 45

The Microsoft® Excel® spreadsheets Constant Velocity Duct Design and Constant Velocity Duct Design Example, available with this book at www.ashrae.org/DuctSyst, are intended for use with the web-based ASHRAE Duct Fitting Database (DFDB; ASHRAE 2016) when designing industrial local exhaust systems with the constant velocity method. Chapter 1 of this design guide provides information on how to use the DFDB, and Chapter 2 offers guidance on how to section a duct system.

HOODS

A hood is a collection device positioned to conduct unwanted contaminants, particles, vapors, gases, smoke, and fumes away from an area. Consult *Industrial Ventilation: A Manual of Recommended Practice for Design* (ACGIH 2019) to select and design hoods for local exhaust systems. The most effective hood uses the minimum exhaust airflow rate to provide maximum contaminant control. Capture effectiveness should be high, but it can be difficult and costly to develop a hood that is 100% efficient. Makeup air supplied by general ventilation to replace exhausted air can sometimes be considered to dilute the contaminants that are not captured by the hood.

Hood Airflow Rates

Use *Industrial Ventilation: A Manual of Recommended Practice for Design* (ACGIH 2019) to calculate hood airflow rates to capture process air contaminants. Note that air quantities are "actual" airflow, i.e., they are based on the air density prevailing in the ventilation system. The velocity into the hood determines the effectiveness of the hood. For example, the same hood used for the same purpose in New Orleans (sea level) and Denver (5000 ft above sea level) should have the same face velocity in both applications (ACGIH 2019, section 6.3.3).

Hoods are typically designed for particulate *control*, not necessarily particulate *collection*. The capture velocity based on actual air is effective in the breathing zone. Particles that settle out need to be cleaned up to prevent reentrainment due to foot traffic and air currents. To capture most particulates, the airflow needs to be corrected for density (by increasing the flow volume and velocity due to elevation changes above sea level).

DUCT SYSTEM

Duct Shape

Round ducts are preferred because they

- offer more uniform velocity to resist settling of material and
- can withstand the higher negative static pressures normally found in exhaust systems.

When design limitations require rectangular or flat oval ducts, the aspect ratio should be as close to unity as possible. That is, both dimensions (height × width or major × minor) need to be as close to the same as possible. Flat oval duct with an aspect ratio of 1 is the same as round duct.

Preferred Duct Fittings

Figures 5-1 through 5-3 show fittings that are preferred, fittings that are acceptable to use, and fittings to avoid due to typically high air velocities and to keep particulates entrained in the airstream. Round elbows with a centerline radius-to-diameter ratio (r/D) from 2 to 2.5 are preferred. Elbows, especially those with large diameters, are often made of seven or more gores. Symmetrical wyes should be 30° to 60°. For converging-flow fittings, a 30° entry angle is recommended to minimize energy losses and abrasion in particulate-handling systems (ASHRAE 2019a). Figure 5-4 shows preferred fan inlet connections, Figure 5-5 shows the design for stackheads, and Figure 5-6 shows the preference of stack designs providing vertical discharge and rain protection. The DFDB (ASHRAE 2016) numbers for these fittings are shown in these figures.

The transport velocity is the minimum velocity necessary to transport particles without them settling in the duct. Table 5-1 lists some generally accepted transport velocities as a function of the nature of the contaminants (ACGIH 2019). The values provided are typically higher than theoretical and experimental values to account for 1) damage to ducts, which increases system resistance and reduces duct velocity and volumetric flow, 2) duct leakage, which tends to decrease velocity in the duct system upstream of the leak, 3) fan wheel corrosion or erosion and/or belt slippage, which could reduce fan volume, and 4) reentrainment of settled particles caused by improper operation of the exhaust system. Design transport velocities can be higher than minimum transport velocities but should never be lower (ASHRAE 2013).

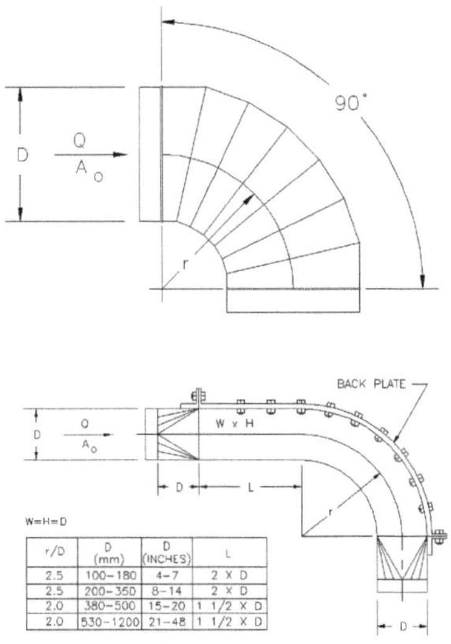

r/D	D (mm)	D (INCHES)	L
2.5	100–180	4–7	2 X D
2.5	200–350	8–14	2 X D
2.0	380–500	15–20	1 1/2 X D
2.0	530–1200	21–48	1 1/2 X D

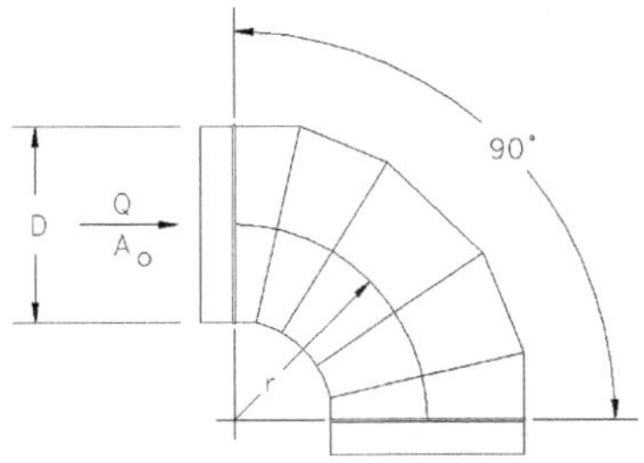

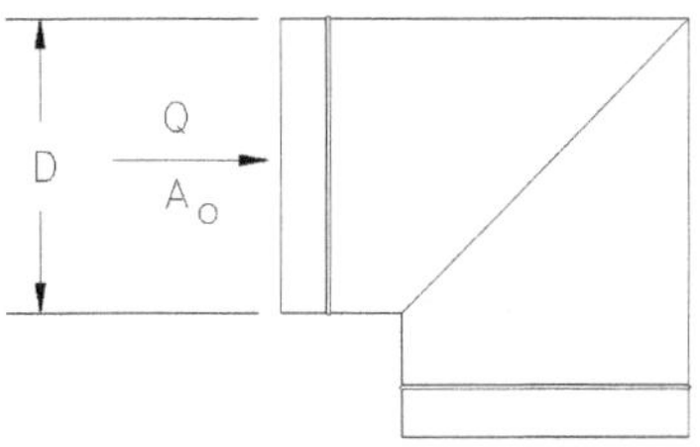

PREFERRED

CD3-10 (7-Gore, 90°, *r*/*D* = 2.5)
CD3-11 (Flat-back, 90°)

ACCEPTABLE

CD3-9 (5-Gore, 90°, *r*/*D* = 1.5)

AVOID

CD3-15 (Mitered, 90°)

(a) Round

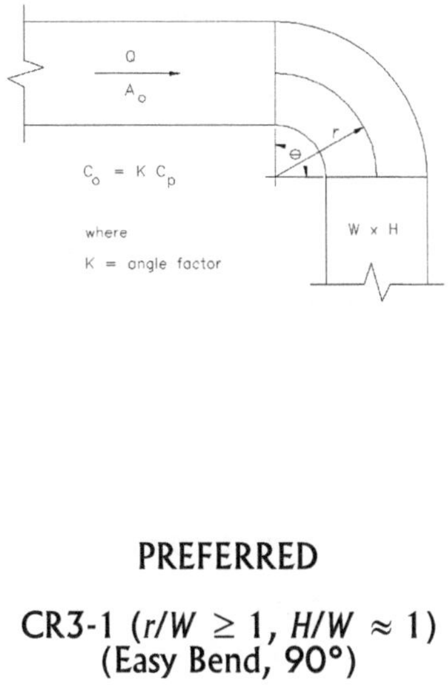

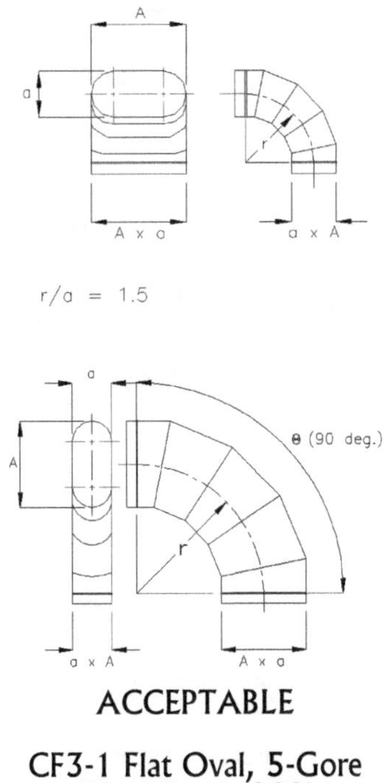

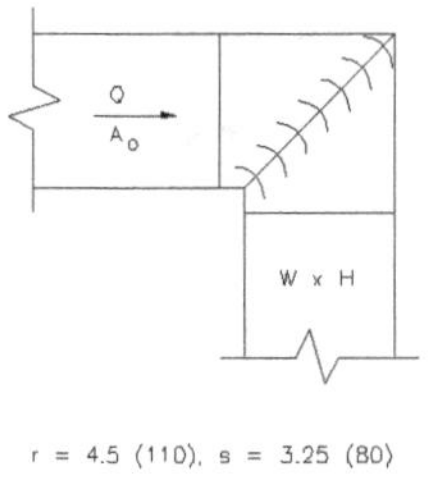

PREFERRED

CR3-1 ($r/W \geq 1$, $H/W \approx 1$)
(Easy Bend, 90°)

ACCEPTABLE

CF3-1 Flat Oval, 5-Gore
(Hard Bend, 90°)
CF3-2 Flat Oval, 5-Gore

AVOID

CR3-6, CR3-9, CR3-12, CR3-14, CR3-15, and CR3-16
90° Mitered Elbow with or without Vanes

(b) Rectangular/Flat Oval

Figure 5-1 Elbows
(Reprinted from ASHRAE 2016)

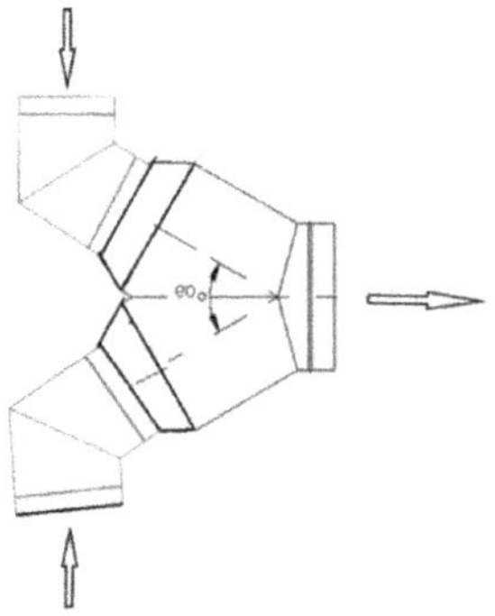

PREFERRED

ED5-9 (60°)
plus CD3-16 (60°)

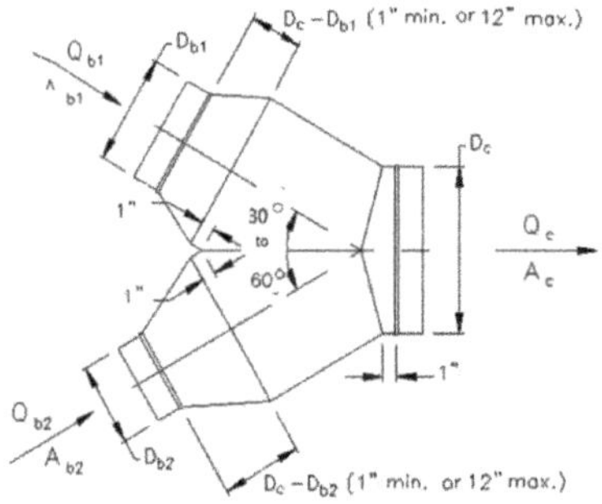

PREFERRED

ED5-9 (60°)

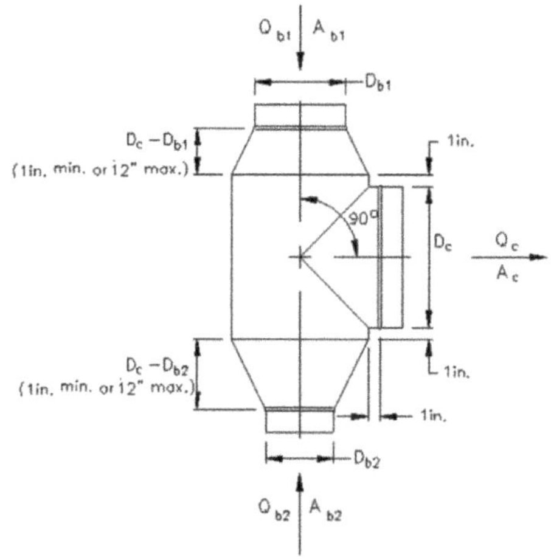

AVOID

ED5-4 Bullhead Tee
without Vanes

Figure 5-2 Symmetrical Wyes (with Equivalent DFDB Fitting Numbers Listed)
(Adapted from Figure 5–21 of Industrial Ventilation: A Manual of Recommended Practice for Design, *30th Edition. © ACGIH 2019. Adapted with permission.)*

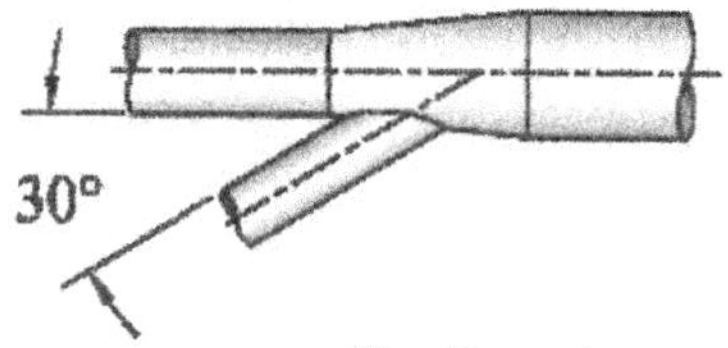

PREFERRED

ED5-1(30° Wye)

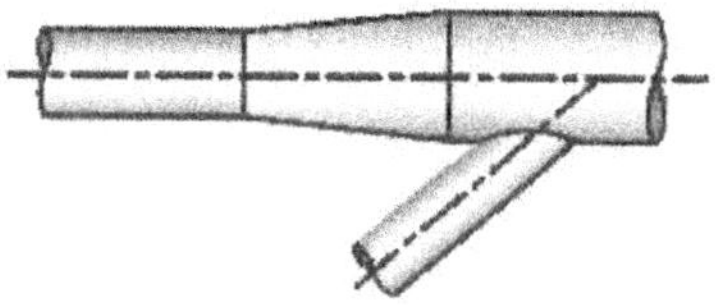

NOT RECOMMENDED

ED4-1 (Wye)
plus ED5-1 (Transition)

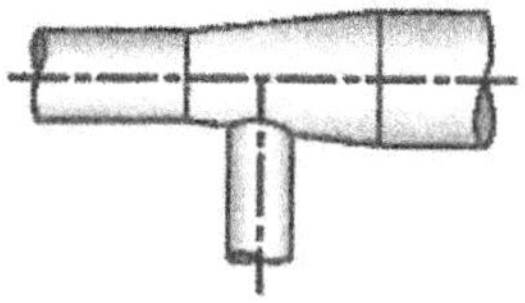

AVOID

ED5-3 (Tee)

Junctions (Laterals and Tees)

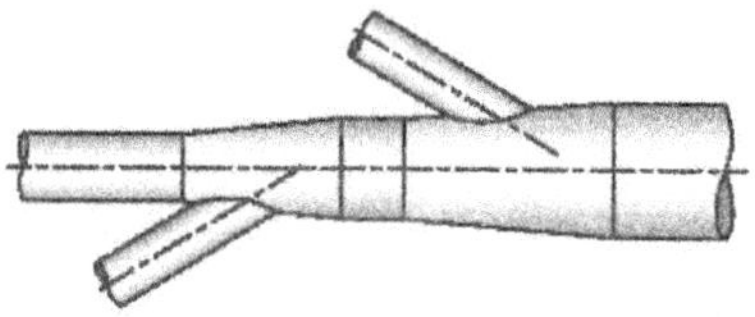

PREFERRED

ED5-1 Wye

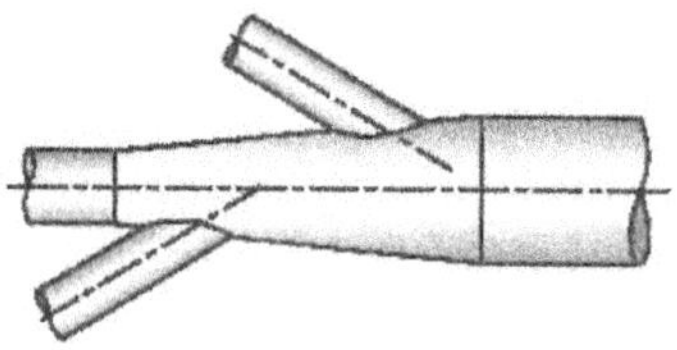

ACCEPTABLE

ED5-1 Wye

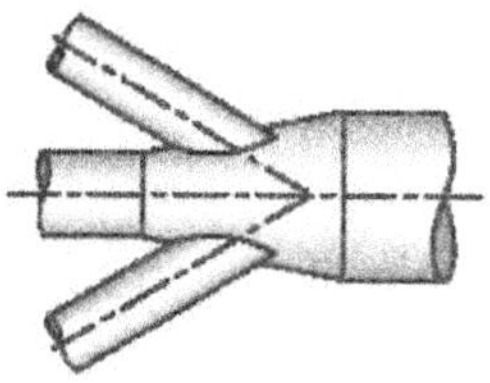

AVOID

ED5-10 Double Wye, 45°
Close Coupled

Branch Entry Arrangements

Figure 5-3 Preferred Branch Entries (Junctions) (with Equivalent DFDB Fitting Numbers Listed)
(Excerpt from Figure 5–10 of Industrial Ventilation: A Manual of Recommended Practice for Design, *30th Edition. © ACGIH 2019. Reprinted with permission.)*

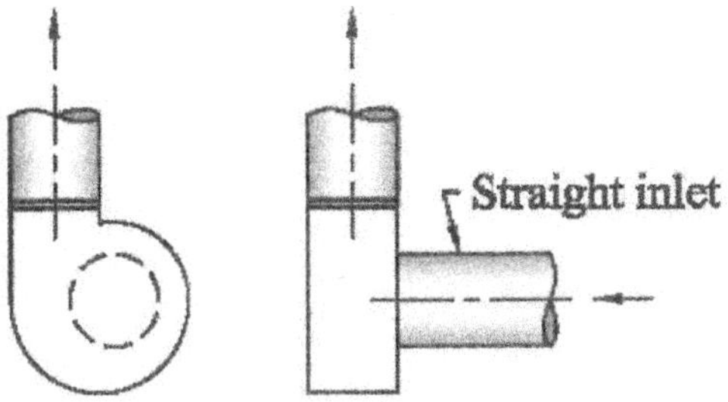

PREFERRED

ED7-2 (Centrifugal, Single Width Single Inlet [SWSI])

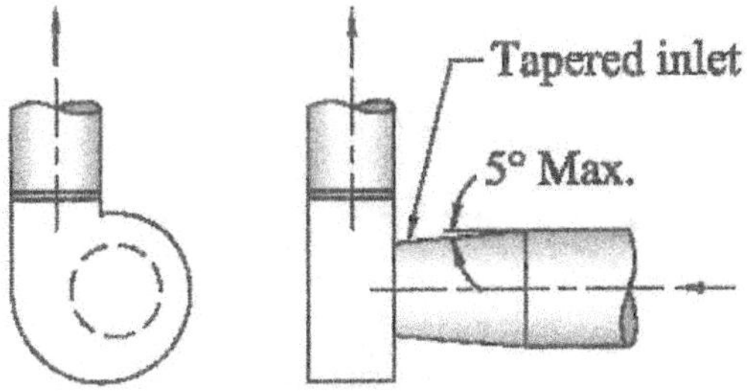

PREFERRED

ED7-2 (Centrifugal, SWSI) with SD4-1 (Transition)

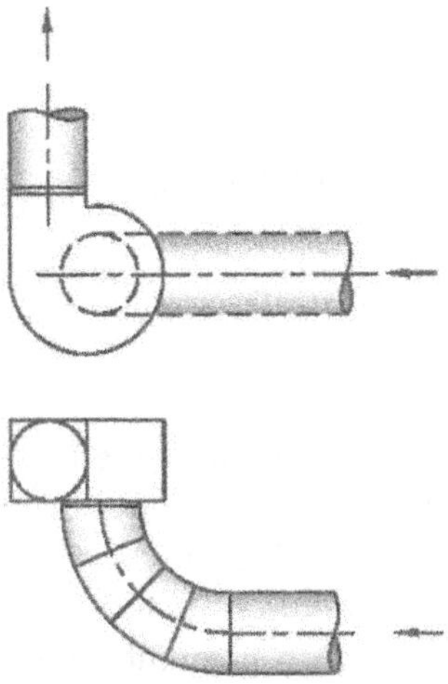

ACCEPTABLE

ED7-2 (Centrifugal, SWSI with 4-Gore Elbow)

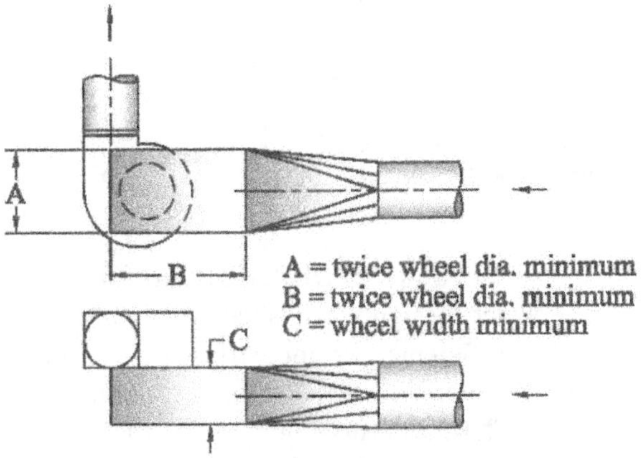

ACCEPTABLE

Figure 5-4 Fan Inlet Connections (with Equivalent DFDB Fitting Numbers Listed)
(Excerpt from Figure 5–23 of Industrial Ventilation: A Manual of Recommended Practice for Design, *30th Edition. © ACGIH 2019. Reprinted with permission.)*

Note that the airflow required in a duct to transport contaminants is not affected by density. The air velocity determines the effectiveness of the duct to transport contaminants, not the mass flow rate (ACGIH 2019, sections 5.3.6 and 6.3.3).

Pressure Losses

The total pressure (the sum of the velocity pressure and the static pressure) indicates the energy of the air flowing in a duct system. As air moves from the inlets through the duct system, the total pressure decreases continually, reaching its minimum value at the fan inlet (see the Total Pressure Grade Line figure in Example 5-4). Losses in total pressure—either friction losses or dynamic losses, as described in Chapter 1—are caused by static and kinetic energy being converted into internal energy as either heat or flow separation. Friction losses are caused by moving air flowing in contact with a fixed boundary, and dynamic losses are produced by turbulence or volume flow rate, direction, shape, or size changes in a duct system (McGill 2003). The designer must determine all of the total pressure losses of the individual components of the exhaust system to determine the fan design to overcome the losses.

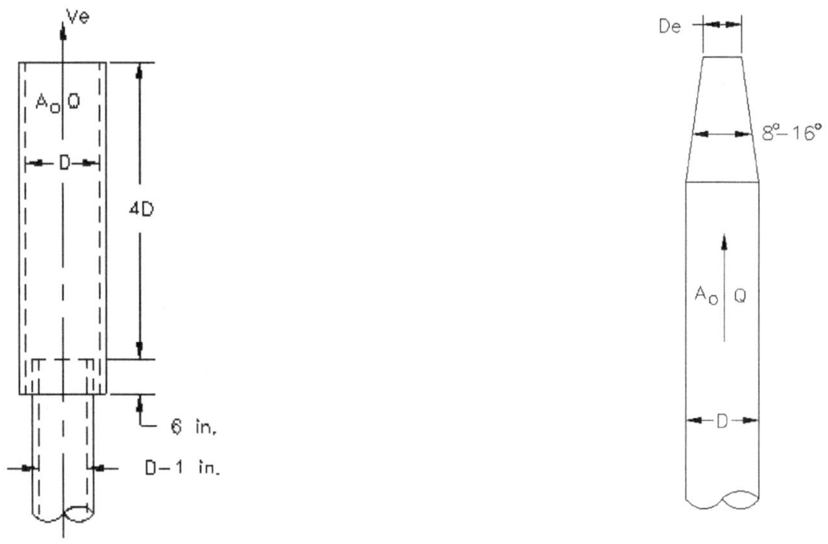

SD2-6 ($D_e/D = 1$) SD2-7 ($D_e/D < 1$)

Figure 5-5 Stackheads
(Reprinted from ASHRAE 2016)

Use the DFDB (ASHRAE 2016) or other loss coefficient tables with the Constant Velocity Duct Design spreadsheet, available with this book at www.ashrae.org/DuctSyst, to get the loss coefficients for industrial local exhaust system fittings, including the entry fitting (hood). Chapter 1 covers the background, nomenclature, and other features of the DFDB. Using the DFDB, it is easy to determine the loss coefficients for both the main (straight-through) and branch of junctions (laterals).

The loss coefficients obtained from the DFDB are total pressure loss coefficients. All calculations in this design guide are total pressure losses. ACGIH calculates static pressure losses; the slot loss coefficient and the hood entry loss coefficient in *Industrial Ventilation: A Manual of Recommended Practice for Design* (ACGIH 2019) are total pressure loss coefficients because they add velocity pressure to the total pressure to get static pressure loss, as shown in Equation 5-1a (adapted from ACGIH 2019, Equation 6.9):

$$SP_h = -[(F_s)(VP_s) + (F_h)(VP_d) + SP_f + VP_d] \tag{5-1a}$$

To get the total pressure loss of the hood, add the duct velocity pressure as shown in Equation 5-1b:

$$TP_h = -[(F_s)(VP_s) + (F_h)(VP_d) + SP_f + VP_d] \tag{5-1b}$$

Example 5-1 shows how to use the DFDB (ASHRAE 2016) to determine main and branch loss coefficients when designing local exhaust systems using the constant velocity method. Example 5-2 shows how hood loss coefficients are calculated.

The DFDB includes canopy hoods, plain openings, flanged openings, and the bell mouth entry (see Table 5-2). Some hoods have multiple pressure losses; for example, a slotted hood with a plenum (see Table 5-2 and Figure 5-7) has multiple losses. This configuration is known as a *com-*

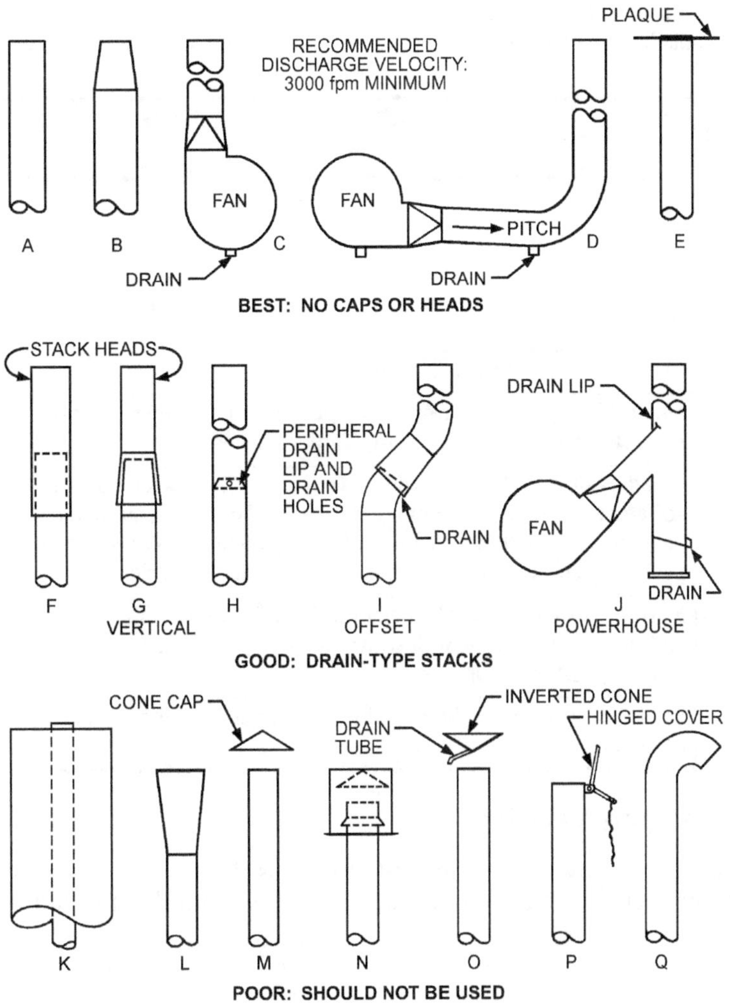

Figure 5-6 Stack Designs Providing Vertical Discharge and Rain Protection
(Reprinted from ASHRAE 2019b, Figure 2)

pound hood. The total pressure loss of this hood is the total pressure loss across the slot plus the total pressure loss from the plenum behind the slot into the adjoining duct section.

For low canopy hoods and for high-temperature processes (see Figure 5-8), the distance from the hood face to the surface of the hot process should be no greater than the diameter of the source or 3 ft, whichever is smaller. Low canopy hoods are simple to design because entrainment of air into the plume and the effects of turbulent cross-drafts are not significant problems. In this case, the diameter of the plume at the hood face is assumed to equal the diameter of the source. *Industrial Ventilation: A Manual of Recommended Practice for Design* (ACGIH 2019) gives Equation 5-4 for the airflow into a low circular receptor hood above a hot process:

$$Q_h = 4.7D_f^{2.33}(T_s - T_o)^{0.41} \tag{5-2}$$

The diameter or side dimensions need be only 1 ft larger than the source. If the source and hood are rectangular, use the following equation:

$$Q_h = 6.2b^{1.33}L(T_s - T_o)^{0.41}$$

Table 5-1 Ranges of Minimum Transport Velocities
(Table 5–1 from Industrial Ventilation: A Manual of Recommended Practice for Design, *30th Edition. © ACGIH 2019. Reprinted with permission)*

Nature of Contaminant	Examples	Design Velocity, fpm
Vapors, gases, smoke		Any desired velocity (economic optimum velocity usually 1000–2000 fpm)
Fumes, metal smokes	Welding	2000–2500 fpm
Fine light dust	Cotton lint, wood flour, litho powder	2500–3000 fpm
Dry dusts and powders	Fine rubber dust, Bakelite molding powder dust, jute lint, cotton dust, shavings (light), soap dust, leather shavings	3000–3500 fpm
Average industrial dust	Grinding dust, buffing lint (dry), wool jute dust (shaker waste), coffee beans, shoe dust, granite dust, silica flour, general material handling, brick cutting, clay dust, foundry (general), limestone dust, packaging and weighing asbestos dust in textile industries	3500–4000 fpm
Heavy dusts	Sawdust (heavy and wet), metal turnings, foundry tumbling barrels and shake-out, sand blast dust, wood blocks, hog waste, brass turnings, cast iron boring dust, lead dust	4000–4500 fpm
Heavy or moist dusts	Lead dusts with small chips, moist cement dust, buffing lint (sticky), quick-lime dust	4500 fpm and up

Duct Size Based on Transport Velocity

Once the designer knows the airflow rate (or mass flow rate and density) and transport velocity, the maximum duct size can be determined by one of the following equations:

$$D = \sqrt{\frac{(144)(4)Q}{\pi V}}$$

$$D = \sqrt{\frac{(144)(4)\dot{m}}{\pi \rho V}}$$

Combining Flows of Different Densities

When two airflows of different densities combine, the continuity equation must be used to calculate the volume flow rate. The mass flow rate $\dot{m}$ in the section with combined flow is the sum of the two branches:

$$\dot{m}_{combined} = \dot{m}_{branch1} + \dot{m}_{branch2} \tag{5-3}$$

The mass flow rate is given by the following:

$$\dot{m} = \rho Q$$

Substituting in Equation 5-2:

$$\rho_{combined} Q_{combined} = \rho_{branch1} Q_{branch1} + \rho_{branch2} Q_{branch2}$$

$$Q_{combined} = \frac{\rho_{branch1} Q_{branch1} + \rho_{branch2} Q_{branch2}}{\rho_{combined}}$$

Example 5-3 shows how to determine a combined flow rate and duct diameters for a section of ductwork.

Example 5-1.
Using the DFDB (ASHRAE 2016), determine the main and branch loss coefficients for a 30° converging wye with D_s = 10 in., D_b = 8 in., D_c = 12 in., Q_s = 2200 cfm, and Q_b = 1400 cfm at a density of 0.062 lb_m/ft^3.

Solution.
See the figure below for the DFDB input/output. The main and branch loss coefficients are –0.03 and 0.12; the main and branch velocity pressures are 0.84 and 0.83 in. of water. Note that the main loss coefficient is negative. This means that the energy stored in the lower-velocity airstream increases because of mixing.

ED5-1 Wye, 30 Degree, Converging (Sepsy 1973)

INPUT		
Diameter (Ds)	in.	10.0
Diameter (Db)	in.	8.0
Diameter (Dc)	in.	12.0
Flow Rate (Qs)	cfm	2200
Flow Rate (Qb)	cfm	1400
Density (RHO)	lbm/ft^3	0.062

Calculate

OUTPUT		
BRANCH		
Velocity (Vb)	fpm	4,011
Vel Pres at Vb (Pvb)	in. wg	0.83
Loss Coefficient (Cb)		0.12
Branch Pressure Loss (Pob)	in. wg	0.10
MAIN		
Velocity (Vs)	fpm	4,034
Velocity (Vc)	fpm	4,584
Vel Pres at Vs (Pvs)	in. wg	0.84
Vel Pres at Vc (Pvc)	in. wg	1.09
Loss Coefficient (Cs)		-0.03
Main Pressure Loss (Pos)	in. wg	-0.03

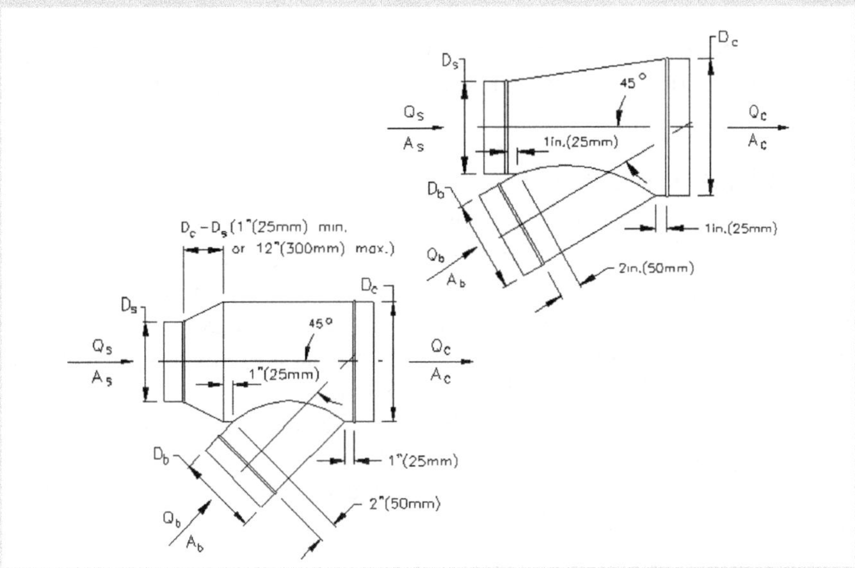

DFDB Output
(Reprinted from ASHRAE 2016, fitting ED5–2)

Example 5-2.

Determine the pressure loss of a multislot hood with a plenum as depicted in Figure 5-7. The multislot hood has four slots, each 1 × 40 in. At the top of the hood is a 90° transition into a 10 in. diameter duct. The volumetric flow required for this hood is 2200 acfm. Assume the air density is 0.075 lb_m/ft^3. The purpose of the slot is to distribute air over the hood face and not influence capture efficiency. Slot velocity should be approximately 2000 fpm to provide air distribution at minimum energy cost; plenum velocities are typically 50% of slot velocity. Higher velocities can cause hot spots (areas of higher-pressure loss contributing to the overall pressure loss) at the face of the hood.

Solution.

The slot velocity and velocity pressure are calculated as follows, where the slot loss coefficient is 1.78:

$$\text{Slot velocity } (V) = \frac{144Q}{A} = \frac{(144)(2200)}{(4)(40)(1)} = 1980 \text{ fpm}$$

$$\text{Slot velocity pressure } (p_v) = \rho\left(\frac{V}{1097}\right)^2 = 0.075\left(\frac{1980}{1097}\right)^2 = 0.24 \text{ in. of water}$$

The plenum to duct loss coefficient is 0.25. The duct velocity and velocity pressure are 4034 fpm and 1.01 in. of water calculated as follows:

$$\text{Duct velocity } (V) = \frac{144Q}{A} = \frac{(144)(2200)}{(\pi)(10^2)/4} = 4034 \text{ fpm}$$

$$\text{Duct velocity pressure } (p_v) = \rho\left(\frac{V}{1097}\right)^2 = 0.075\left(\frac{4034}{1097}\right)^2 = 1.01 \text{ in. of water}$$

$$\begin{aligned}\Delta p_{t,\,slot} &= C_{o,\,slot}p_{v,\,slot} + C_{o,\,canopy}p_{v,\,duct} \\ &= (1.78)(0.24) + (0.25)(1.01) \\ &= 0.68 \text{ in. of water}\end{aligned}$$

System Balancing

In a multiple-hood system it is necessary to provide a means of distributing airflow between the branches of a junction either by balanced design or by the use of blast gates (ASHRAE 2016, fitting CD9-2).

Balance by Adding Resistance

In general, the balance of flows in multibranch duct systems can be improved by adding resistance to one or more branches. For example, it is possible to change the resistance of either path through section 3 or 4 to attain total pressure balance at the junction. For paths conveying particulates, additional resistance is added to the path of lower resistance by decreasing the duct size, thus maintaining the particulate transport velocity. For paths not conveying particulates, the path with the greater resistance can be increased to attain balance.

When the difference in total pressure between the branches is less than 20%, airflow balance can be obtained by increasing the airflow through the path with the lower resistance. This change in flow rate can be calculated by Equation 5-3 because pressure losses vary with velocity pressure, and therefore as the square of the flow rate:

$$Q_c = Q_L\left(\frac{\Delta p_H}{\Delta p_L}\right)^{1/2} \qquad (5\text{-}4)$$

Table 5-2 DFDB Hood Fitting Numbers Used to Determine Loss Coefficients

(Adapted from Figure 6–39, Industrial Ventilation: A Manual of Recommended Practice for Design, *30th Edition. © ACGIH 2019. Adapted with permission.)*

Hood Type	Description	DFDB Fitting Number
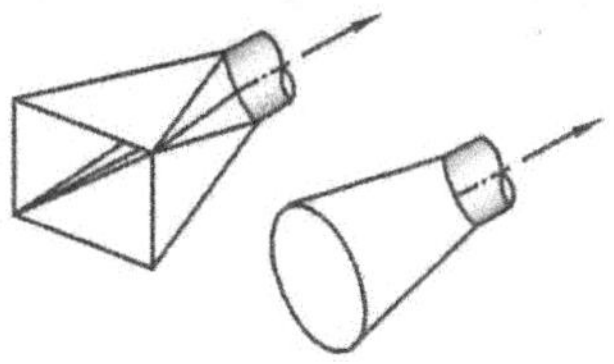	Canopy hood	ED10-1 ED10-2
	Plain opening	ED1-1
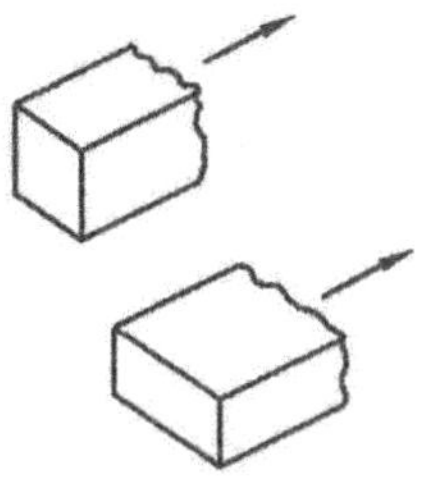		ER1-1
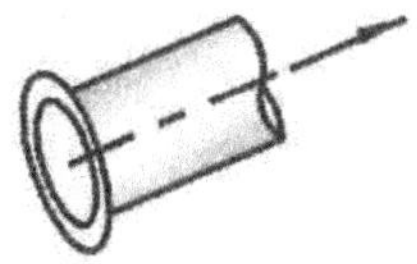	Flanged opening	ED1-1
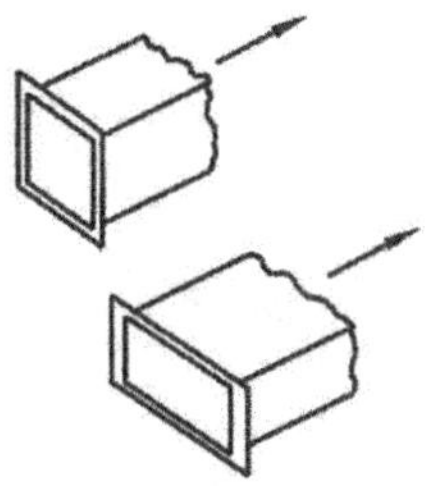		ER1-1
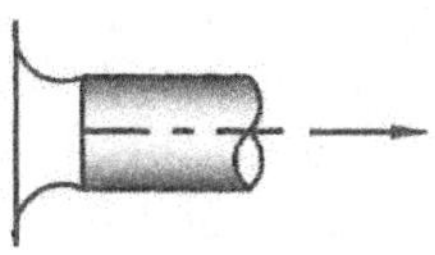	Bellmouth inlet	ED1-2
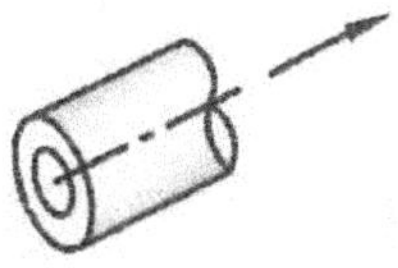	Orifice	ED1-7
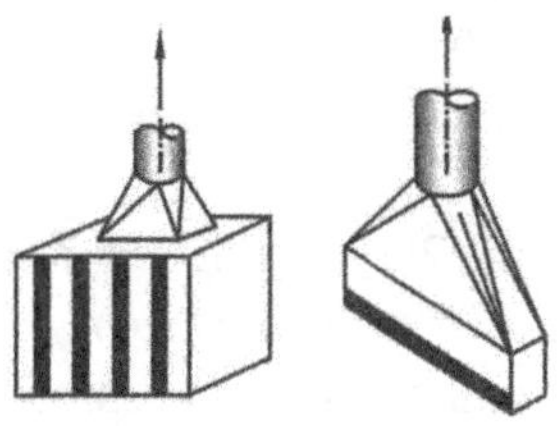	Slot hood	N/A—Consult ACGIH (2019, Chapter 13)

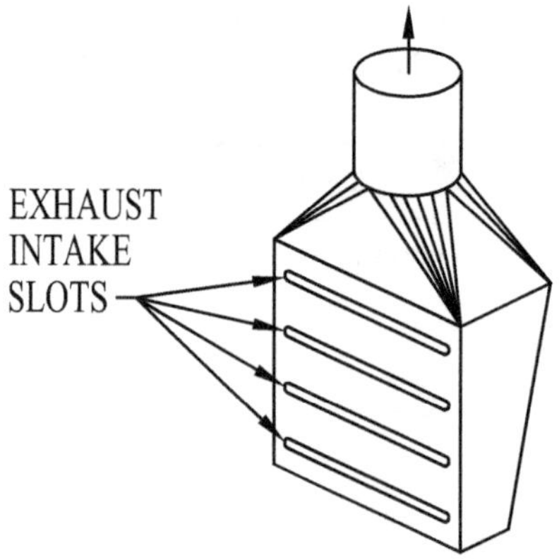

Figure 5-7 Multislot Non-Enclosing Hood
(Adapted from Figure VS-55-10 of Industrial Ventilation: A Manual of Recommended Practice for Design, *30th Edition. © ACGIH 2019. Adapted with permission.)*

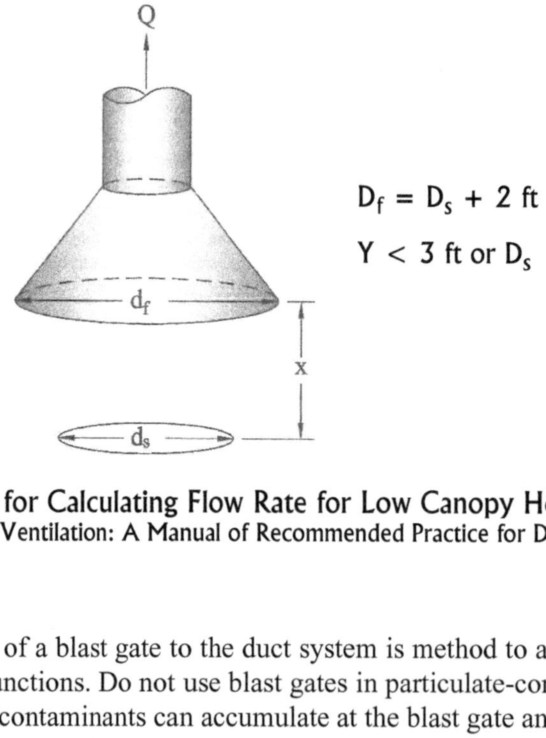

Figure 5-8 Geometry for Calculating Flow Rate for Low Canopy Hoods
(Figure 13-27-1 from Industrial Ventilation: A Manual of Recommended Practice for Design, *30th Edition. © ACGIH 2019. Reprinted with permission.)*

The addition of a blast gate to the duct system is method to add resistance to attain total pressure balance at junctions. Do not use blast gates in particulate-conveying sections of local exhaust systems because contaminants can accumulate at the blast gate and wear away the blade.

Example 5-4 shows how to design a duct system with given parameters and Example 5-5 shows how density is calculated in sections 3, 4 and 5 of the system shown in Example 5-4.

MIXING TWO AIRSTREAMS

Many exhaust duct systems involve mixing two airstreams. One stream of contaminants may be at one temperature and the other stream at another. At the junction of two airstreams, as in a lateral fitting, the upstream and branch meet to form the downstream side of the fitting as shown in Figure 5-9.

The equations in this section are applicable for temperatures up to 200°F. For temperatures greater than 200°F use the calculation technique supplied by ACGIH (2019, Section 9.6). ACGIH uses psychrometric charts that go up to 1500°F. For airstreams not exceeding 200°F, the spreadsheet

Example 5-3.

Determine the combined flow rate in section 5 (Q_5) and the duct diameters for sections 3 and 5 of Example 5-4 (see the table below). The figure below shows the components and section numbers of this example. Do not calculate the duct size for section 4 because there is no minimum velocity requirement.

Section	Temperature, °F	Q, acfm	$\dot{m}$, lb_m/min	ρ, lb_m/ft^3	Minimum Velocity, fpm
3	90	1300	93.0	0.072	4000
4	200	815	46.8	0.057	Any
5	130	2122	139.8	0.066	4000

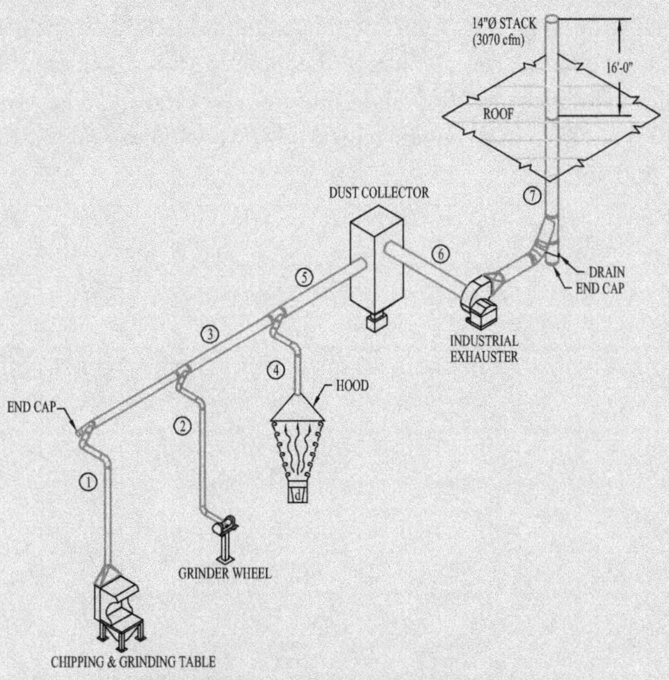

Industrial Local Exhaust System
(Adapted from ASHRAE 2017, Figure 30)

Solution.

The actual airflow rate in section 5 is calculated as follows:

$$Q_5 = \frac{0.072 \times 1300 + 0.057 \times 815}{0.066} = 2122 \text{ acfm}$$

$$D_3 = \sqrt{\frac{(144)(4)(Q_3)}{\pi V_3}} = \sqrt{\frac{(144)(4)(1300)}{(3.14159)(4000)}} = 7.7 \text{ in.}$$

$$D_5 = \sqrt{\frac{(144)(4)(Q_5)}{\pi V_5}} = \sqrt{\frac{(144)(4)(2122)}{(3.14159)(4000)}} = 9.9 \text{ in.}$$

Example 5-4.

Design a system as shown in the Industrial Local Exhaust System figure below with the hood design and requirements summarized in the table that follows. The hood air quantities are supported by the example problems noted in the table. System elevation is sea level. Ambient air is 90°F and 50% rh. Size ducts to 1/2 in. increments. Duct material is galvanized steel. Section numbers are per the schematic shown in the System Schematic with Section Numbers figure of this example. The dust collector has a total pressure loss of 3 in. of water.

The building is one story, and the design wind velocity is 20 mph. For the stack, use design J of Figure 5-6 for complete rain protection. The stack height is determined by calculations from Chapter 46, "Building Air Intake and Exhaust Design," of *ASHRAE Handbook—HVAC Applications* (2019b) and is 16 ft above the roof. This height is based on minimized stack downwash; therefore the stack discharge velocity must exceed 1.5 times the design wind velocity.

The fan outlet diffuser is $H_1 = 9\ 3/4$ in., $W_1 = 13$ in., and $D_o = 11$ in. The airflow requirement of the chipping and grinding table hood (see its figure in this example) is 800 acfm, with 600 acfm above the table and 200 acfm below the table. The slot above the table is 36 × 2.4 in. and that below is 36 × 4 in. The upper and lower slot velocities are 1000 and 200 fpm, respectively. The grinding wheel hood (see its figure in this example) is 18 in. in diameter with a 3 in. wheel width and has a good enclosure. The exhaust requirement is 500 acfm.

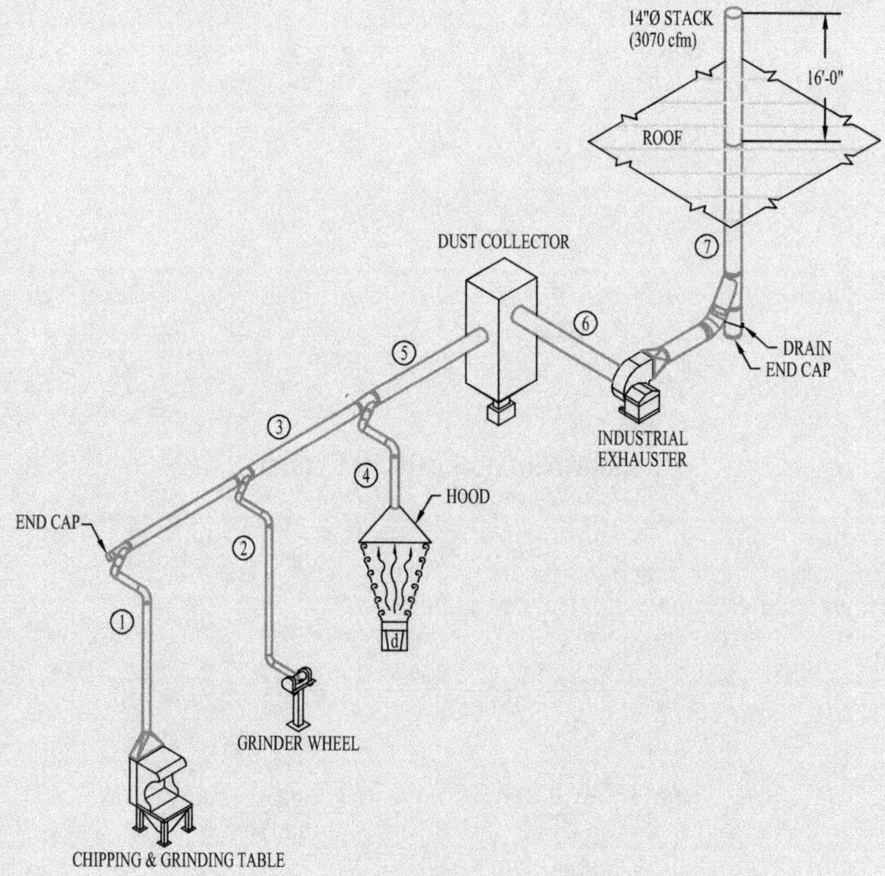

Industrial Local Exhaust System

(Adapted from ASHRAE 2017, Figure 30)

Summary of Airflow and Air Density Calculations and Design Criteria

Section	Hood	Airflow, acfm	Temperature, Relative Humidity	Air Density, lb_m/ft^3	Minimum Duct Velocity, fpm (Table 5-1)
1	See the Chipping and Grinding Table figure in this example	800 (See Example 5-5)	90°F, 50% rh	0.072	3500
2	See the Grinding Wheel Hood figure in this example	500 (See Example 5-4)	90°F, 50% rh	0.072	4000
3	—	1300	90°F, 50% rh	0.072	4000
4	See Figure 5-8	815 (See Example 5-5)	200°F, 14% rh	0.051 (See Example 5-5)	Any
5	—	2122	130°F, 36% rh (See Example 5-5)	0.0665 (See Example 5-5)	4000
6	—	2122	130°F, 36% rh	0.065	Any
7	—	2122	130°F, 36% rh	0.065	Any
—	Exit	—	—	—	2640

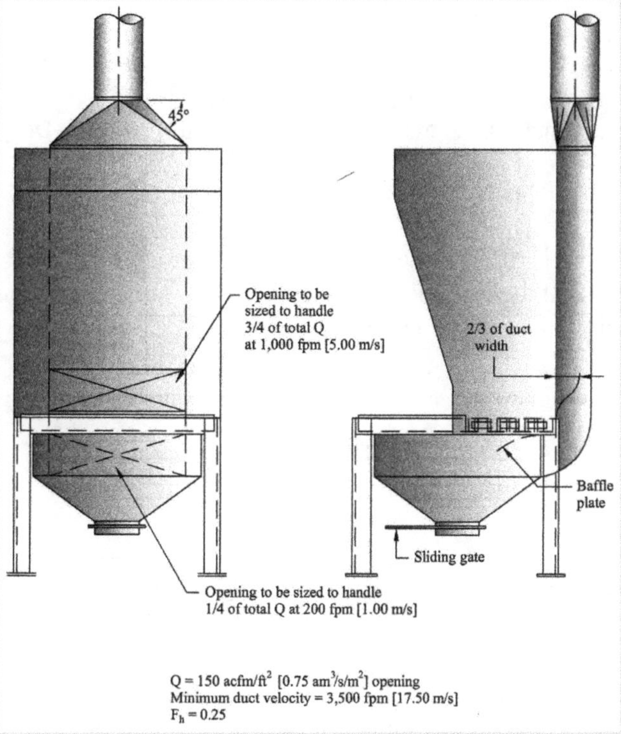

Chipping and Grinding Table

(Figure VS-80-19 from Industrial Ventilation: A Manual of Recommended Practice for Design, *30th Edition. © ACGIH 2019. Reprinted with permission.)*

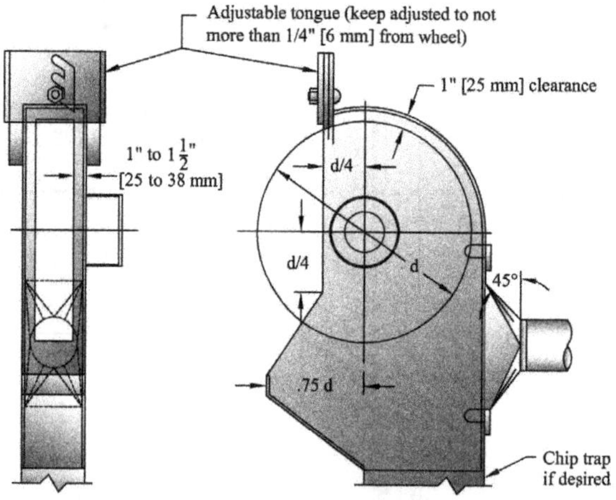

EXHAUST FLOW RATES, acfm [am³/s]

Wheel dia. inches	[mm]	Wheel width inches	[mm]	Good enclosure* acfm	[am³/s]	Poor enclosure acfm	[am³/s]
Up to 5	[Up to 125]	1	[25]	220	[0.11]	220	[0.11]
5 to 10	[125 to 250]	1.5	[38]	220	[0.11]	300	[0.15]
10 to 14	[250 to 350]	2	[50]	300	[0.15]	500	[0.25]
14 to 16	[350 to 400]	2	[50]	390	[0.20]	610	[0.31]
16 to 20	[400 to 500]	3	[75]	500	[0.25]	740	[0.37]
20 to 24	[500 to 600]	4	[100]	610	[0.31]	880	[0.44]
24 to 30	[600 to 750]	5	[125]	880	[0.44]	1,200	[0.60]
30 to 36	[750 to 900]	6	[150]	1,200	[0.60]	1,600	[0.80]

* No more than 25% of wheel exposed.
Minimum duct velocity = 4,000 fpm [20.00 m/s]
F_h = 0.65 for straight take-off
F_h = 0.40 for tapered take-off

Grinding Wheel Hood

(Figure VS-80-11 from **Industrial Ventilation: A Manual of Recommended Practice for Design,** *30th Edition. © ACGIH 2019. Reprinted with permission.)*

Solution.

The figure below is a schematic of the system with section numbers assigned according to the guidance laid out in Chapter 1. Note that the loss coefficient of the bellmouth connection to the collector (ER2-1) is referenced to the velocity pressure of the duct (C1).

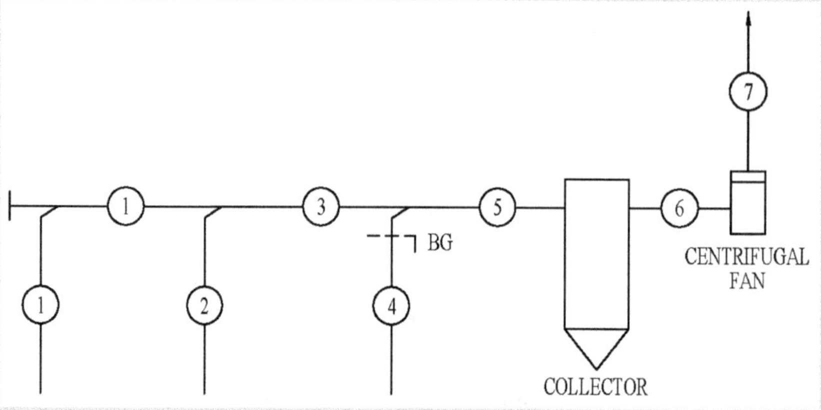

System Schematic with Section Number

The iterative solution to maintain velocity and minimum total pressure difference between sections 1 and 2 is summarized in the following table. The solution is shown in the white row.

D_1, in.	D_2, in.	D_3, in.	$\Delta p_{t,1}$, in. of water	$\Delta p_{t,2}$, in. of water	$\Delta p_{t,1} - \Delta p_{t,2}$, in. of water
6.5	4.5	7.5	1.44	2.60	−1.16
6	4.5	7.5	2.25	2.60	−0.35
5.5	4.5	7.5	3.62	2.49	+1.13

Solution for Sections 1 and 2

The iterative solution to maintain velocity and minimum total pressure difference between sections 2 and 3 and section 4 is summarized in the following table. The solution is shown in the white row.

D_3, in.	D_4, in.	D_5, in.	$\Delta p_{t,2} + \Delta p_{t,3}$, in. of water	$\Delta p_{t,4}$, in. of water	$\Delta p_{t,2+3} - \Delta p_{t,4}$, in. of water
7.5	5.5	9.5	2.60 + 0.77 = 3.37	1.67	+1.70
7.5	5.0	9.5	2.60 + 0.58 = 3.18	2.90	+0.28
7.5	4.5	9.5	2.60 + 0.33 = 2.93	5.01	−2.08

Solution for Sections 3 and 4

Section 6, the duct between the collector and the fan inlet, is 18 in. round at a friction rate of 0.07 in. of water per 100 ft of duct. In section 7, to minimize downwash, the stack discharge velocity must exceed 2640 fpm, 1.5 times the design wind velocity (20 mph) as stated in the problem definition. Therefore, the stack discharge is 12 in. round, and the stack discharge velocity is 2693 fpm.

Example 5-5 shows how density is calculated in sections 3, 4 and 5. However the DFDB (ASHRAE 2016) was used to calculated density by entering the air properties (0.072, 0.051, and 0.065 lb_m/ft^3) using the equation below. Rounding of the values during manual calculation may cause slight differences in the results compared with those from the DFDB calculations.

The total pressure grade line is shown in the figure below. The fan rating is 2122 cfm at 0.065 lb_m/ft^3 density and 9.0 in. of water total pressure. Balance at junction 3-4-5 can be obtained by using a blast gate in section 4 (a nonparticulate section). Balance at junction 1-2-3 can be attained by adjustment of the airflow when the system is in operation. No blast gates should be located in sections 1 or 2 because particulates are being transported there.

The airflow adjustment for sections 1 and 2 can be calculated by Equation 5-3 as follows:

$$Q_2 = Q_1\left(\frac{\Delta p_2}{\Delta p_1}\right)^{1/2} = 800\left(\frac{2.60}{2.25}\right)^{1/2} = 860 \text{ acfm}$$

However, in this case the calculations were not modified because this 7% correction is less than the accuracy of the calculations. (See the Uncertainty and Error Considerations section of Chapter 2 for an analysis of accuracy in duct design.)

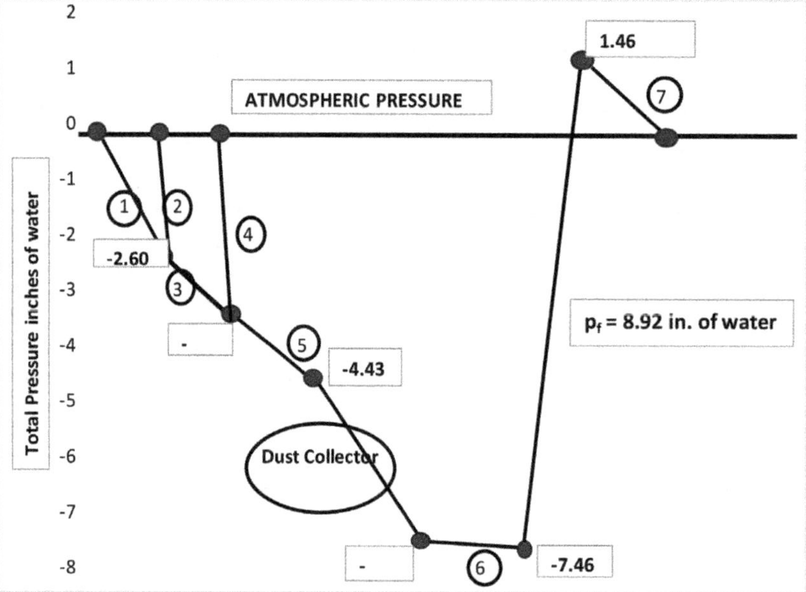

Total Pressure Grade Line

(Excerpt from the Constant Velocity Duct Design Example spreadsheet available with this book online; adapted from ASHRAE 2017, Figure 32)

1	2	3	4	5	6	7	8	9	10	11	12	13	14	15	16
Section	Parent	Brother	Fitting Description	ASHRAE Fitting Code	Air Qty, acfm	Temp., °F	Density, lbm/ft3	Minimum Velocity, fpm (Table 6-1)	Maximum Duct Dia., D_e, in.	Duct Size D (W x H, A x a), in.	Actual Velocity, fpm	Duct Length, ft	Velocity Pressure, pv, in. of water	Loss Coefficient C0	Δp_t, in. of water
				Source											
				DFDB	Drawing	Drawing	DFDB	Table 6-1	Eq. 6-2	Round Down	DFDB	Drawing	DFDB	Table 6-2, ACGIH, DFDB	Σ
1	3	2	Hood: ACGIH, VS-80-19 [a]	Appendix C, Figure C-1	800	90	0.072							0.25	
			Duct	CD11-1				3500	6.47	6	4074	38	[Round Dia]		1.45
			90° Flat-back Elbow	CD3-11										0.08	
			Capped wye with elbow: Db=6 in.. Dc=6	ED5-6										0.61	
			Wye, main: Ds=6 in.. Dc=7.5 in., Db=4.5in.	ED5-1										-0.13	
													0.99	0.81	0.80
Section 1 Total															2.25
2	3	1	Hood: ACGIH (2010), VS-80-11[b] (Tapered Takeoff)	Appendix C, Figure C-2	500	90	0.072							0.40	
			Duct	CD11-1				4000	4.79	4.5	4527	20	[Round Dia]		1.33
			90° Flat-back Elbow	CD3-11										0.09	
			90° Flat-back Elbow	CD3-11										0.09	
			45° Die Stamped Elbow	CD3-3										0.12	
			Wye, branch: Ds=6 in.. Dc=7.5 in., Db=4.5in.	ED5-1										0.34	
													1.22	1.04	1.27
Section 2 Total															2.60
3	5	4	Duct	CD11-1	1300	90	0.072	4000	7.72	7.5	4237	30	[Round Dia]		0.94
			Wye, main: Dc=9.5 in., Ds=7.5 in., Db=5.0 in.	ED5-1										-0.34	
													1.07	-0.34	-0.36
Section 3 Total															0.58
4	5	3	Hood: ACGIH (2010). Figure 13-27-1	90° ED10-1 Appendix C, Figure C-3	815	200	0.051							0.16	
			Duct	CD11-1				TBD	--	5	5977	14	[Round Dia]		1.04
			90° Die Stamped Elbow	CD3-1										0.16	
			45° Die Stamped Elbow	CD3-3										0.10	
			Damper, blast gate,	CD9-2, h=5 (fully open)										0.00	
			Wye, branch: Dc=9.5 in., Ds=7.5 in., Db=5.0	ED5-1										0.60	
													1.52	1.02	1.55
Section 4 Total															2.59

Constant Velocity Duct Design Example

(Excerpt from the Constant Velocity Duct Design Example spreadsheet available with this book online)

Example 5-5.
A hot bath at 200°F has a circular canopy hood at 2 ft above the hot source. The ambient temperature is 90°F. The diameter of the hot source (D_s) is 2 ft. What is the total airflow into the hood?

Solution.
Using Equation 5-4, the total airflow into the hood is found to be 815 acfm:

$$D_f = D_s + 2 = 4 \text{ ft}$$

$$Q_h = 4.7 D_f^{2.33}(T_s - T_o)^{0.41} = (4.7)(4)^{2.33}(200 - 90)^{0.41} = 816 \text{ acfm}$$

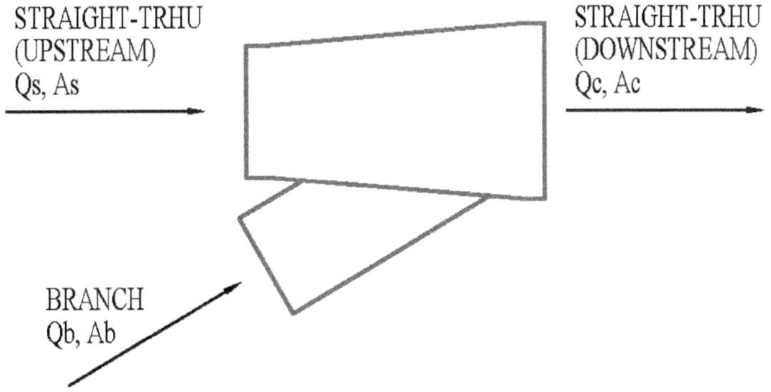

Figure 5-9 Mixing at a Junction

tool ASHRAE Duct Calculator, available with this book at www.ashrae.org/DuctSyst, can be used for calculating mixed-air psychrometric properties. If the airstreams are at the same density conditions, the volume flow rates can be directly added together. However, if two or more processes are being exhausted in the same system and the conditions of the processes are not the same, the mass flow rates of the airstreams must be calculated and combined using Equation 5-5:

$$\dot{m}_c = \dot{m}_s + \dot{m}_b \tag{5-5}$$

The mass flow of hoods is calculated using Equation 5-6 because generally the air quantity and density of each airstream are known:

$$\dot{m}_h = \rho_h Q_h \tag{5-6}$$

The density of moist or dry air can be obtained by using the DFDB (ASHRAE 2016) or from *ASHRAE Handbook—Fundamentals* (2017). If there are significant quantities of particulate in the duct system (>20 grains per standard cfm), this addition to the airstream should be addressed. However, such situations (the field is called *material conveying*) are beyond the scope of this guide. Note that 20 grains per standard cfm particulate represent less than 0.4% of the air mass rate; this represents significant amounts of air to move a small amount of particulate (ACGIH 2019, section 9.6).

The actual airflow in the common (Q_c) section of combined flows is calculated by Equation 5-7, where the combined airstream density (ρ_c) is calculated by Equations 5-8 through 5-14.

$$Q_c = \frac{\dot{m}_c}{\rho_c} \tag{5-7}$$

PSYCHROMETRIC EQUATIONS TO CALCULATE DENSITY OF ANY AIRSTREAM

Barometric pressure, p_b, is calculated from Equation 5-8:

$$p_b = 14.696(1 - 6.8754 \times 10^{-6} Z)^{5.2559} \quad (5\text{-}8)$$

Knowing barometric pressure (p_b), temperature (T), and relative humidity (ϕ), the humidity ratio (W), enthalpy (h), specific volume (ν), and density (ρ) of air can be calculated using Equations 5-8 through 5-14. The "Moist Air Transport Properties" tab of the spreadsheet tool ASHRAE Duct Calculator, available with this book at www.ashrae.org/DuctSyst, can be used to calculate these properties.

$$p_{ws} = \text{function}(T) \quad (5\text{-}9)$$

$$p_w = \phi p_{ws} \quad (5\text{-}10)$$

$$W = 0.622\frac{p_w}{(p_b - p_w)} \quad (5\text{-}11)$$

$$h = 0.24T + W(1061 + 0.444T) \quad (5\text{-}12)$$

$$\nu = \frac{0.370(T_c + 460)(1 + 1.6079 W_c)}{p_b} \quad (5\text{-}13)$$

$$\rho = \frac{1 + W}{\nu} \quad (5\text{-}14)$$

The degree of saturation μ is as follows, where W and W_s are evaluated at the same temperature and pressure:

$$\mu = \left.\frac{W}{W_s}\right|_{T,p} \quad (5\text{-}15)$$

Relative humidity (ϕ) is calculated as follows:

$$\phi = \frac{\mu}{1 - (1 - \mu)(p_{ws}/p_b)} \quad (5\text{-}16)$$

PSYCHROMETRIC EQUATIONS TO CALCULATE PROPERTIES OF COMBINED AIRSTREAMS

The humidity ratio of the combined airstream can be calculated from Equation 5-17:

$$W_c = \frac{\dot{m}_s W_s + \dot{m}_b W_b}{\dot{m}_c + \dot{m}_b} \quad (5\text{-}17)$$

The humidity ratios of the common section and branch, W_s and W_b, respectively, are calculated by Equation 5-10, and $\dot{m}_b$ is calculated by Equation 5-6.

The temperature downstream of the junction is determined using Equation 5-18:

$$T_c = \frac{h_c - 1061\,W_c}{0.24 + 0.444\,W_c} \tag{5-18}$$

where

$$h_c = \frac{\dot{m}_s h_s + \dot{m}_b h_b}{\dot{m}_c + \dot{m}_h} \tag{5-19}$$

where h_s and h_b are calculated by Equation 5-12.

Example 5-6 shows how to calculate the temperature, relative humidity, density, and airflow when two fluid streams are mixed.

Example 5-6.

Calculate the temperature, relative humidity, density, and airflow at cross-section "c" of the lateral tee in the local exhaust system depicted in the figure below. The system is at sea level and the ambient temperature is 90°F. Section 3 is 1300 acfm at 90°F and 50% rh. Section 4 is 815 acfm at 200°F with a moisture content of approximately 0.08 lb of water per lb of dry air. What are the airflow, temperature, density, and relative humidity at section 5 (mixture of sections 3 and 4)? (Note: manually calculated results may not exactly match those in the solution because of differences in rounding.)

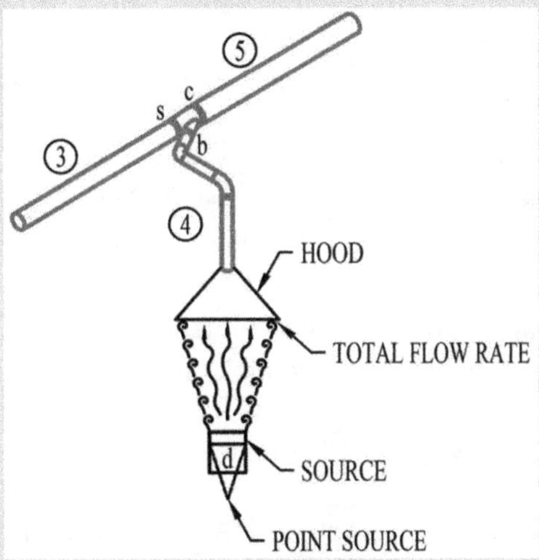

Local Exhaust System

(Adapted from Figure VS-55-03b of Industrial Ventilation: A Manual of Recommended Practice for Design, *30th Edition. © ACGIH 2019. Adapted with permission.)*

Solution.

The barometric pressure at sea level is 14.696 psia calculated using Equation 5-8.

For psychrometric conditions at section 3: Using ASHRAE Duct Calculator (available with this book online), the partial pressure of water at 90°F when saturated is found to be $p_{ws} = 0.6989$ psia. The partial pressure of water vapor at 90°F and 50% rh is $p_{ws} = 0.5 \times 0.6989 = 0.3494$ psia (Equation 5-9). The humidity ratio of moist air at section 3, calculated by Equation 5-11, is

$$W_s = 0.622\frac{p_{ws}}{(p_b - p_{ws})} = 0.622\left(\frac{0.3494}{14.696 - 0.3494}\right) = 0.1515\ (\mathrm{lb_w/lb_{da}})$$

The enthalpy at section 3, calculated by Equation 5-12, is

$$h_s = 0.24T_s + W_s(1061 + 0.444T_s) = (0.24)(90) + 0.01515(1061 + 0.444 \times 90) = 38.3 \text{ Btu/lb}_{da}$$

The specific volume at section 3, calculated by Equation 5-13, is

$$v = \frac{0.370(T_s + 460)[1 + 1.6079W_s]}{p_b} = \frac{0.370(90 + 460)[1 + 1.6079(0.01515)]}{14.696} = 14.2 \text{ ft}^3/\text{lb}_{da}$$

The density of moist air at section 3, calculated by Equation 5-14, is

$$\rho_s = \frac{(1 + W_s)}{v_s} = \frac{1 + 0.01515}{14.2} = 0.072 \text{ lb}_m/\text{ft}^3$$

The mass flow rate at section 3, calculated by Equation 5-6, is

$$\dot{m}_s = \rho_s Q_s = 93.0 \text{ lb}_m/\text{min}$$

For psychrometric conditions at section 4: The humidity ratio at section 4 is approximately 0.08 lb_w per lb_{da}. The relative humidity is calculated as follows:

$$\mu = \left.\frac{W}{W_s}\right|_{T,p} = \left.\frac{0.08}{2.2720}\right|_{200°\text{F},\ 14.696\text{ psia}} = 0.0352$$

$$\phi = \frac{\mu}{1 - (1 - \mu)(p_{ws}/p_b)} = \frac{0.0352}{1 - (1 - 0.0352)(11.5374/14.696)} = 0.14$$

where the partial pressure of water at 200°F when saturated is $p_{ws,b}$ = 11.5374 psia. (For this calculation use ASHRAE Duct Calculator, available with this book online.)

$$p_{w,b} = \phi p_{ws,b} = (0.14)(11.5374) = 1.6748 \text{ psia}$$

The humidity ratio of moist air at section 4, calculated by Equation 5-11, is

$$W_b = 0.622\frac{p_{w,b}}{(p_b - p_{w,b})} = 0.622\frac{1.6748}{14.696 - 1.6748} = 0.0800 \text{ lb}_w/\text{lb}_{da}$$

The enthalpy at section 4, calculated by Equation 5-12, is

$$h_b = 0.24T_b + W_b(1061 + 0.444T_b) = (0.24)(200) + 0.0800(1061 + 0.444 \times 200) = 140 \text{ Btu/lb}_{da}$$

The specific volume at section 4, calculated by Equation 5-13, is

$$v_b = \frac{0.370(T_b + 460)[1 + 1.6079W_b]}{p_b} = \frac{0.370(200 + 460)[1 + 1.6079(0.0800)]}{14.696} = 18.8 \text{ ft}^3/\text{lb}_{da}$$

The density of moist air at section 4, calculated by Equation 5-14, is

$$\rho_b = \frac{1 + W_b}{v_b} = \frac{1 + 0.0800}{18.8} = 0.058 \text{ lb}_\text{m}\text{ft}^3$$

The mass flow rate at section 4, calculated by Equation 5-6, is

$$\dot{m}_b = \rho_b Q_b = (0.058)(815) = 46.9 \text{ lb}_\text{m}/\text{min}$$

For psychrometric conditions at section 5: The humidity ratio of moist air at section 5, calculated by Equation 5-17, is

$$W_c = \frac{\dot{m}_s W_s + \dot{m}_b W_b}{\dot{m}_s + \dot{m}_b} = \frac{(93.0)(0.01515) + (46.9)(0.0800)}{93.0 + 46.9} = 0.03688 \text{ lb}_\text{w}/\text{lb}_\text{da}$$

The enthalpy at section 5, calculated by Equation 5-19, is

$$h_c = \frac{\dot{m}_s h_s + \dot{m}_b h_b}{\dot{m}_s + \dot{m}_b} = \frac{(93.0)(38.3) + (46.9)(140)}{93.0 + 46.9} = 72.4 \text{ Btu/lb}_\text{da}$$

The temperature at section 5, calculated by Equation 5-18, is

$$T_c = \frac{h_c - 1061 W_c}{0.24 + 0.444 W_c} = \frac{71.9 - (1061 \times 0.03688)}{0.24 + 0.444 \times 0.03688} = 129.6°\text{F}$$

The specific volume at section 5, calculated by Equation 5-13, is

$$v_c = \frac{0.370(T_c + 460)[1 + 1.6079 W_c]}{p_b} = \frac{0.370(129.6 + 460)[1 + (1.6079)(0.03688)]}{14.696} = 15.72 \text{ ft}^3/\text{lb}_\text{da}$$

The density of moist air at section 5, calculated by Equation 5-14, is

$$\rho_c = \frac{1 + W_c}{v_c} = \frac{1 + 0.03688}{15.72} = 0.066 \text{ lb}_\text{m}/\text{ft}^3$$

The mass flow rate at section 5, calculated by Equation 5-5, is

$$\dot{m}_c = \dot{m}_s + \dot{m}_b = 93.0 + 46.9 = 139.9 \text{ lb}_\text{m}/\text{min}$$

Finally, the airflow at section 5, calculated by Equation 5-7, is

$$Q_c = \frac{\dot{m}_c}{\rho_c} = \frac{(1)(139.9)}{(1)(0.066)} = 2122 \text{ cfm}$$

By arithmetic addition, the air quantity at section 5 is 2115 cfm; however, because of the density change in section 5, the actual airflow is 2122 acfm. In this specific case the airflow difference is negligible. In many instances the effects of air density variation can be ignored, but that influence may be significant when higher temperatures are encountered.

The relative humidity ϕ at section 5 is 37% calculated as follows:

$$p_{ws,\,c} = 2.2022 \text{ psia}$$

$$W_{ws,\,c} = 0.622\frac{p_{ws,\,c}}{(p_b - p_{ws,\,c})} = 0.622\frac{2.2022}{14.696 - 2.2022} = 0.1096 \text{ lb}_w/\text{lb}_{da}$$

$$\mu = \left.\frac{W_c}{W_{ws,\,c}}\right|_{129.6°\text{F},\ 14.696 \text{ psia}} = \frac{0.03688}{0.1096} = 0.34$$

$$\phi = \frac{\mu}{1-(1-\mu)(p_{ws,\,c}/p_b)} = \frac{0.34}{1-(1-0.34)(2.2022/14.696)} = 0.37$$

NOMENCLATURE

A	=	cross-sectional area of duct, in.^{2}
A_o	=	stackhead cross section, in.^{2}
b	=	width of rectangular hood, ft
C_o	=	loss coefficient, dimensionless
D	=	diameter, in.
D_b	=	diameter of branch section, in.
D_c	=	diameter of common section, in.
D_e	=	stackhead exit diameter, in.
D_f	=	diameter of plume, ft
D_o	=	outlet diameter, in.
D_s	=	diameter of hot source, ft, or diameter of straight section, in.
F_h	=	hood entry to duct loss, in. of water
F_s	=	slot loss coefficient, dimensionless
H_1	=	fan outlet diffuser height, in.
h	=	enthalpy of moist air, Btu/lb$_{da}$
h_b	=	enthalpy upstream of junction (branch), Btu/lb$_{da}$
h_c	=	enthalpy downstream of junction (common), Btu/lb$_{da}$
h_s	=	enthalpy upstream of junction (straight-through section), Btu/lb$_{da}$
L	=	length of rectangular hood, ft
$\dot{m}$	=	mass flow rate, lb$_m$/min
$\dot{m}_b$	=	mass flow rate of branch (Figure 5-9), lb$_m$/min
$\dot{m}_c$	=	mass flow rate of common section (Figure 5-9), lb$_m$/min
$\dot{m}_h$	=	hood mass flow rate, lb$_m$/min
$\dot{m}_s$	=	mass flow rate of straight-through section (Figure 5-9), lb$_m$/min
p	=	pressure, psia
p_b	=	barometric pressure, psia
p_t	=	total pressure, psia
p_v	=	velocity pressure, in. of water
p_w	=	partial pressure of water vapor in moist air, psia

$p_{w,b}$ = partial pressure of water vapor in moist air (branch), psia
p_{ws} = pressure of saturated water, psia
$p_{ws,b}$ = pressure of saturated water (branch), psia
$p_{ws,c}$ = pressure of saturated water (common), psia
Q = airflow, cfm
Q_c = airflow rate of combined flows (common), acfm
Q_c = airflow rate of combined flows (straight-through section), acfm
Q_C = corrected airflow, acfm
Q_h = total airflow into the hood, acfm
Q_L = airflow through low-resistance duct segment, acfm
r = elbow turning radius, in.
SP_f = hood filter static pressure loss, in. of water
SP_h = hood static pressure loss, in. of water
T = temperature, °F
T_b = temperature downstream of junction (branch), °F
T_c = temperature downstream of junction (common), °F
T_s = temperature downstream of junction (straight-through section), °F
$T_s - T_o$ = temperature difference between hot-source surface and ambient air, °F
TP_h = hood total pressure loss, in. of water
V = velocity, fpm
V_e = stackhead exit velocity, fpm
VP_d = duct velocity pressure, in. of water
VP_s = slot or opening velocity pressure, in. of water
W = humidity ratio of moist air, lb_w/lb_{da}
W_1 = fan outlet diffuser width, in.
W_b = humidity ratio upstream of junction (branch section), lb_w/lb_{da}
W_c = humidity ratio of moist air downstream of junction (common), lb_w/lb_{da}
W_s = humidity ratio upstream of junction (straight-through section), lb_w/lb_{da}
$W_{ws,c}$ = humidity ratio of saturated water (common), lb_w/lb_{da}
Z = altitude, ft

Symbols

Δp_H = pressure loss in high-pressure duct section, in. of water
Δp_L = pressure loss in low-pressure duct section, in. of water
μ = degree of saturation
ν = specific volume of moist air, ft^3/lb_{da}
ν_b = specific volume of moist air (branch), ft^3/lb_{da}
ν_c = specific volume of moist air (common), ft^3/lb_{da}
ν_s = specific volume of moist air (straight-through section), ft^3/lb_{da}
ρ = air density, lb_m/ft^3
ρ_b = density of combined airstream (branch), lb_m/ft^3
ρ_c = density of combined airstream (common), lb_m/ft^3
ρ_h = hood airstream density, lb_m/ft^3
ρ_s = density of combined airstream (straight-through section), lb_m/ft^3
ϕ = relative humidity

REFERENCES

ACGIH. 2019. *Industrial ventilation: A manual of recommended practice for design*, 30th ed. Cincinnati, OH: American Conference of Governmental Industrial Hygienists.

ASHRAE. 2013. Chapter 2, Air contaminants and control. In *ASHRAE pocket guide for air conditioning, heating, ventilation, refrigeration*. Peachtree Corners, GA: ASHRAE.

ASHRAE. 2016. ASHRAE Duct Fitting Database, ver. 6.00.05 (online). Peachtree Corners, GA: ASHRAE. www.ashrae.org/technical-resources/bookstore/duct-fitting-database.

ASHRAE. 2017. Chapter 21, Duct design. In *ASHRAE handbook—Fundamentals*. Peachtree Corners, GA: ASHRAE.

ASHRAE. 2019a. Chapter 33, Industrial local exhaust systems. *ASHRAE handbook—HVAC applications*. Peachtree Corners, GA: ASHRAE.

ASHRAE. 2019b. Chapter 46, Building air intake and exhaust design. *ASHRAE handbook—HVAC applications*. Peachtree Corners, GA: ASHRAE.

McGill. 2003. Chapter 1, Airflow fundamentals for supply duct systems. In *Duct System Design Guide*. Groveport, OH: McGill AirFlow Corporation.

6

Fan-Duct System Interaction

FAN TYPES AND CHARACTERISTICS

According to Chapter 21, "Fans," of *ASHRAE Handbook—HVAC Systems and Equipment* (2016b), there are generally two main categories of fans: centrifugal and axial. Table 6-1 summarizes the impeller and housing designs, performance curves, characteristics, and common uses of each type of fan. Airfoil and backward-curved centrifugal fans are generally used for ducted HVAC systems; on occasion, vane axial fans may be used. Radial centrifugal fans are typically used for industrial local exhaust systems. Plug fans are unhoused centrifugal fan impellers that are used as circulators in some industrial applications (e.g., heat-treating ovens). In this case, there is no duct connection to the fan because it simply circulates the air within the oven. Plenum fans are unhoused centrifugal fan impellers located in a plenum chamber or an air-handling unit (AHU) with the fan inlet connected to an inlet duct or an adjacent section of the AHU. Other types of fans are also referenced in Table 6-1.

FAN STATIC PRESSURE

It is often not clear to many designers that the addition of the word "fan" to modify the terms *total pressure*, *static pressure*, or *velocity pressure* changes these definitions. It is almost a universal misconception among non-fan professionals that the static pressure rise across a fan is equivalent to the fan static pressure. This is often exacerbated by the common application of the term *static pressure* in fan catalogs when *fan static pressure* is meant. The definitions of fan total pressure, fan static pressure, and fan velocity pressure are as follows:

- Fan total pressure is "the difference between the total pressure at the fan outlet and the total pressure at the fan inlet" (AMCA 2016, p. 2). This is P_t in Figure 6-1.
- Fan static pressure is "the difference between the fan total pressure and the fan dynamic (velocity) pressure. Therefore, it is the difference between static pressure at the fan outlet and total pressure at the fan inlet" (AMCA 2016, pp. 2, 6). This is P_s in Figure 6-1, calculated by Equation 6-1:

$$P_t = P_s = p_{v,o} \tag{6-1}$$

- Fan velocity pressure is the pressure corresponding to the average air velocity at the fan outlet. It equals the fan total pressure minus fan static pressures, as can be seen by rearranging the terms in Equation 6-1.

Table 6-1 Types of Fans

(Adapted from ASHRAE 2017, Table 1)

Centrifugal—Backward Inclined (includes AF, BC, and BI)	
Impeller Design	**Housing Design**
Blades are inclined away from direction of rotation and can be single thickness flat (BI), single thickness curved (BC), or airfoil (AF) contour. Deep blades allow efficient expansion within blade passages. Air leaves impeller at velocity less than tip speed. For given duty, has highest speed of centrifugal fan designs. 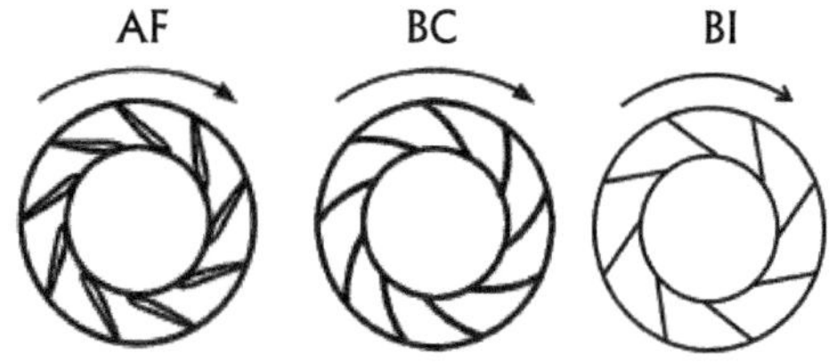	Single or double inlet scroll design for efficient conversion of tangential velocity pressure to static pressure. Maximum efficiency requires close clearance and alignment between impeller and inlet. 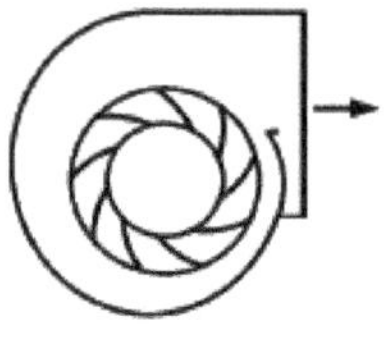
Performance Characteristics	**Applications**
Highest efficiency of all centrifugal fan designs with peak efficiencies occurring at 50 to 60% of wide open volume. Fan has a non-overloading characteristic, which means power reaches maximum near peak efficiency and becomes lower, or self-limiting, toward free delivery. Airfoil blades are most efficient, followed by curved and then flat blades. 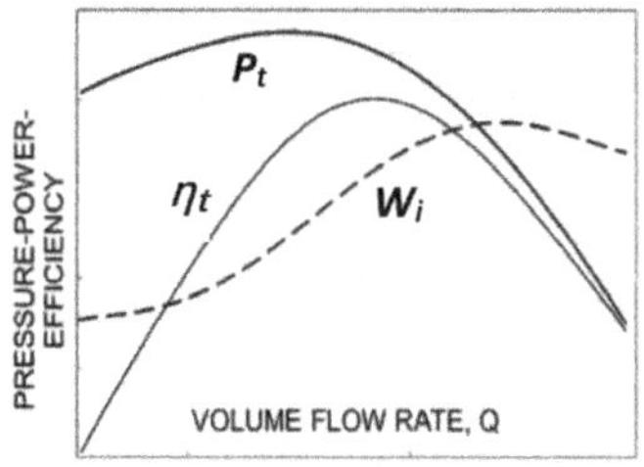	General heating, ventilating, and air conditioning. Used in ducted applications covering a large range of pressures. Airfoil blades can be used for clean air industrial operations. For industrial applications where the environment may corrode or erode airfoil blades, consider using single thickness blades instead.
Centrifugal—Radial and Radial Tipped	
Impeller Design	**Housing Design**
Blades are either fully radial (R) or backward inclined with a radial curve at outer edge (RT). Fully radial blades can have back plate and shroud, back plate only, or neither (open). Radial tipped blades normally have back plate and shroud. 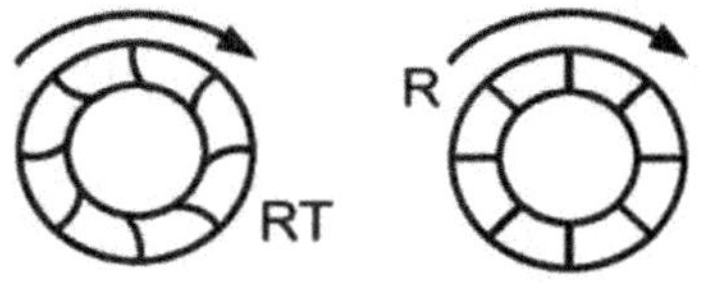	Single or double inlet scroll similar to other centrifugal fan designs. Fit between impeller and inlet not as critical as for backward inclined fans. 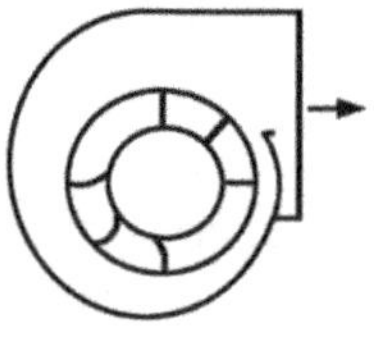
Performance Characteristics	**Applications**
Higher pressure characteristics than airfoil and backward curved fans. Power rises continually to free delivery, which is an overloading characteristic. Radial tipped blades are slightly more efficient than straight blades. Shrouded impellers are more efficient than non-shrouded. 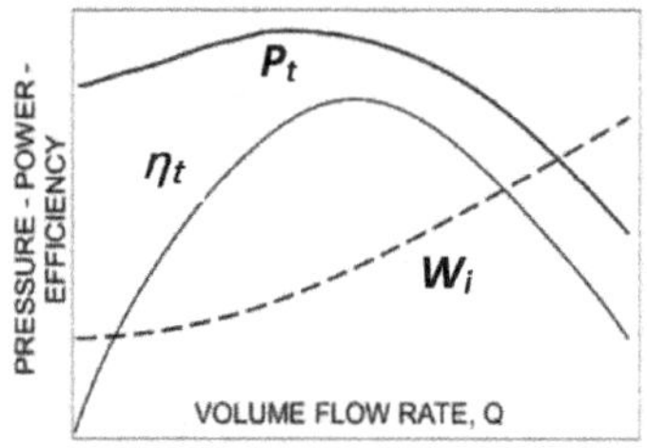	Primarily for material handling in industrial plants where high duct velocities are required to keep materials airborne. Also for some high pressure industrial requirements. Rugged impeller is simple to repair in the field. Choice of impeller type generally depends on materials being transported. Impeller sometimes coated with special material.

Table 6-1 Types of Fans *(Continued)*
(Adapted from ASHRAE 2017, Table 1)

Centrifugal—Forward Curved	
Impeller Design Large number of thin curved blades with outer edge tipped toward direction of rotation. Air leaves impeller with a high tangential velocity. Relies on scroll housing to convert this velocity pressure to static pressure. For given duty, has lowest speed of centrifugal fan designs. 	**Housing Design** Single or double inlet scroll shaped design necessary for conversion of tangential velocity pressure to static pressure. Fit between impeller and inlet not as critical as for backward inclined fans. 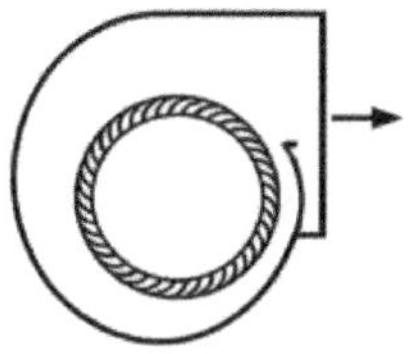
Performance Characteristics Pressure curve less steep than that of backward-inclined fans. Curve dips to left of peak pressure. Highest efficiency occurs at 40 to 50% of wide open volume. Operate fan to right of peak pressure. Power rises continually to free delivery which is an overloading characteristic. 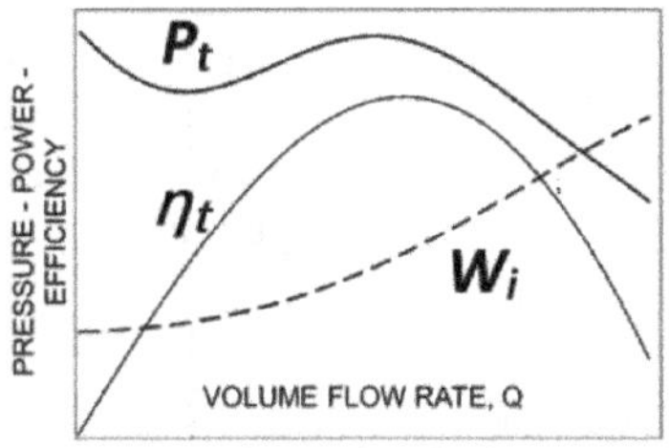	**Applications** Primarily for low to medium pressure HVAC applications, such as residential furnaces, central station units, and packaged air conditioners. Slower speed and lower tonal noise characteristics allow them to be used closer to occupied spaces. A possible exception is central station air handling units because of low frequency noise that is more difficult to attenuate.
Centrifugal—Plenum/Plug	
Impeller Design Single inlet centrifugal impellers can be airfoil, backward curved, or backward inclined. Mixed flow or radial impellers can also be used for specific applications. 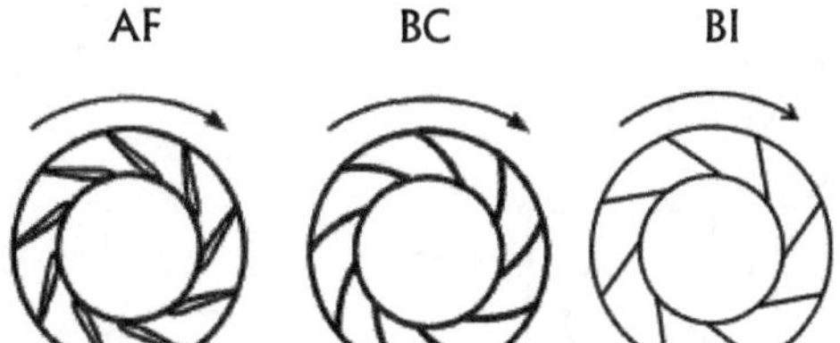	**Chamber Design** No integral housing. The equivalent of a housing, or plenum chamber (dashed line), depends on the application. Due to the lack of a housing, all pressure development occurs in the impeller. The components of the drive system for the plug fan are located outside the airstream. 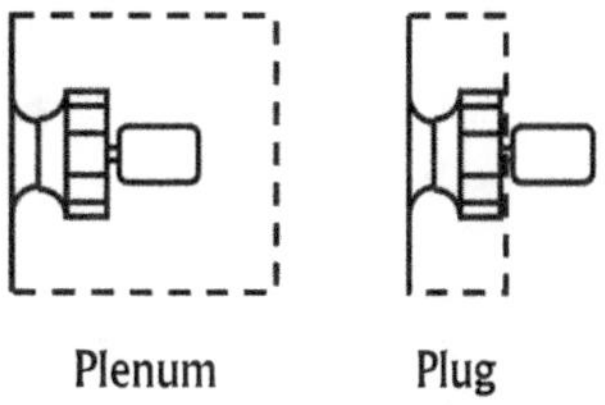
Performance Characteristics Similar to housed airfoil/backward curved/backward inclined fans but slightly less efficient due to the lack of conversion of kinetic energy in the discharge airstream. Since plenum and plug fans do not have usable outlet velocity pressure, they should always be selected based on fan static pressure. Both fans are susceptible to performance degradation caused by poor inlet installation and insufficient clearance to surrounding walls. 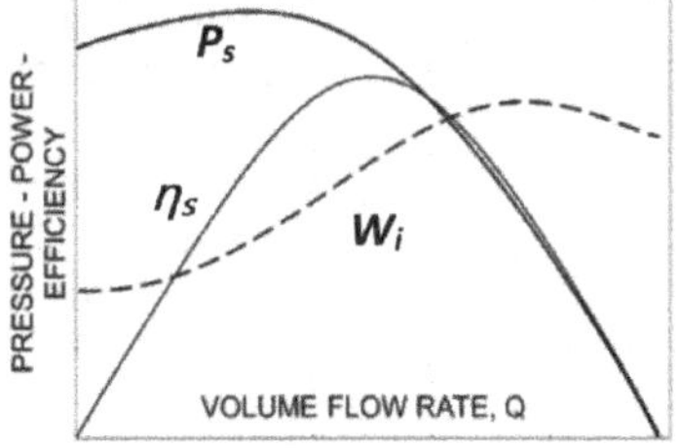	**Applications** Plenum fans are used for HVAC equipment such as air handlers. Advantages include flexibility of equipment discharge and potential for smaller equipment footprint. Multiple plenum fans in parallel (fan arrays) can be used to further reduce the axial length of air handling equipment. Plug fans are commonly used for high temperature process applications.

Table 6-1 Types of Fans *(Continued)*

(Adapted from ASHRAE 2017, Table 1)

Axial—Tube Axial

Impeller Design	Housing Design
Usually has 4 to 8 blades with airfoil or single thickness cross section. Blades may have fixed or adjustable pitch. Hub is usually less than half the fan tip diameter. 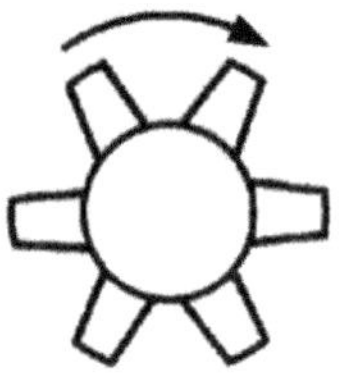	Cylindrical tube with close clearance to blade tips. Free inlet applications should use a bell mouth inlet to minimize entrance losses. 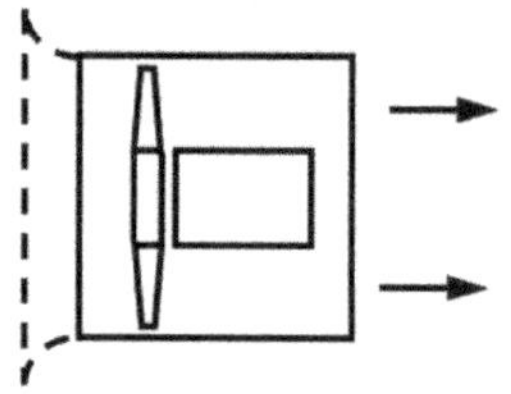
Performance Characteristics	**Applications**
High flow rate, medium pressure capabilities. Pressure curve can dip to left of peak pressure at higher blade pitches. Avoid operating fan in this stall region. Discharge pattern circular and airstream rotates or swirls downstream of fan. 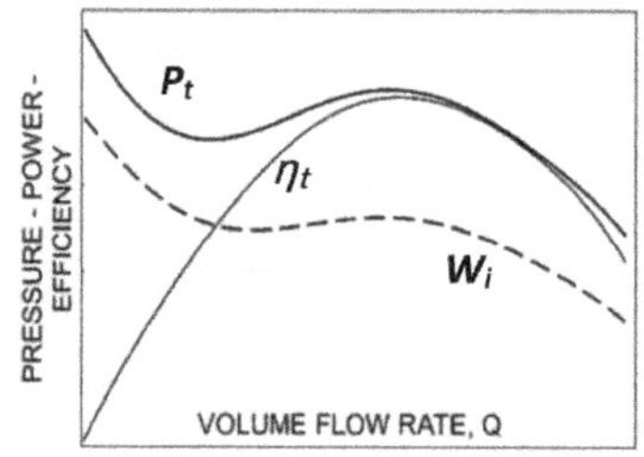	Low and medium pressure ducted HVAC applications where straight through flow and compact installation are required. Used in some industrial applications, such as drying ovens, paint spray booths, and fume exhausts.

Axial—Vane Axial

Impeller Design	Housing Design
Typically has more blades than tube axial fans. Blades may have fixed, adjustable, or controllable pitch. Hub is usually greater than half fan tip diameter. Most efficient designs have airfoil blades. 	Cylindrical tube with close clearance to blade tips. Guide vanes upstream or downstream from impeller increase pressure capability and efficiency. 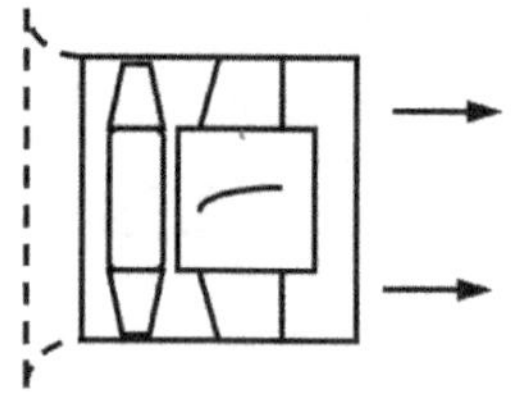
Performance Characteristics	**Applications**
High pressure characteristics with medium volume flow capabilities. Pressure curve can dip to left of peak pressure. Avoid operating fan in this stall region. Guide vanes correct circular motion imparted by impeller which improves pressure characteristics and efficiency of fan. 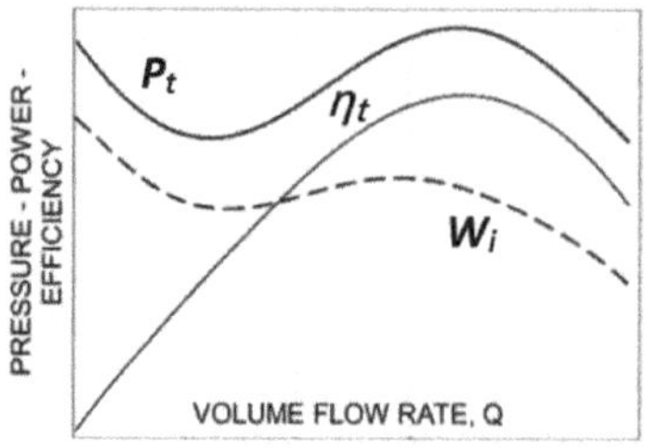	General HVAC systems in medium and high pressure applications where straight through flow and compact installation are required. Has swirl free downstream airflow. Used in industrial applications in place of tube axial fans. More compact than centrifugal fans for same duty.

Table 6-1 Types of Fans *(Continued)*

(Adapted from ASHRAE 2017, Table 1)

Mixed Flow (MF)

Impeller Design

Combination of axial and centrifugal characteristics. Contoured back or hub; airfoil, curved, or straight blades, either shrouded or not. Airflow through impeller has both radial and axial components.

Housing Design

The majority of mixed flow fans have a tubular housing for use in ducted applications and include outlet turning vanes.

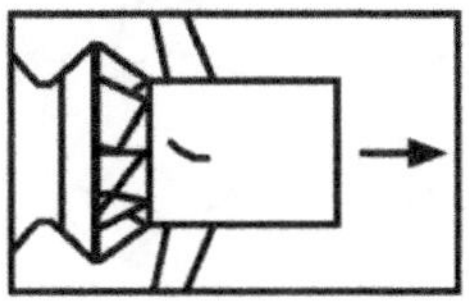

Performance Characteristics

Characteristic pressure curve between axial fans and centrifugal fans. Higher pressure than axial fans and higher volume flow than centrifugal fans.

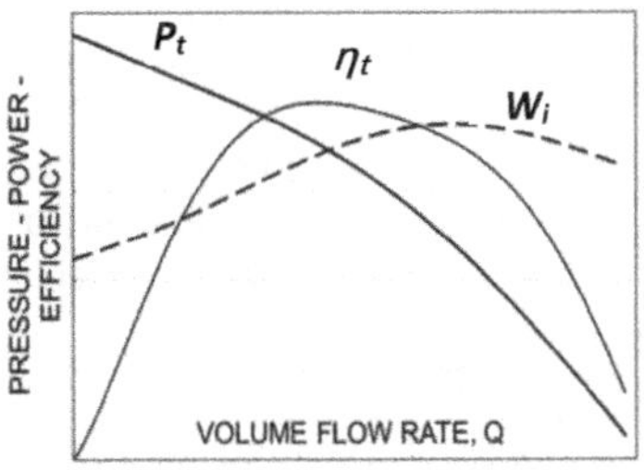

Applications

Similar HVAC applications to centrifugal fans or in applications where an axial fan cannot generate sufficient pressure rise.

Inline Centrifugal

Impeller Design

Single inlet centrifugal impellers can be airfoil, backward curved, or backward inclined.

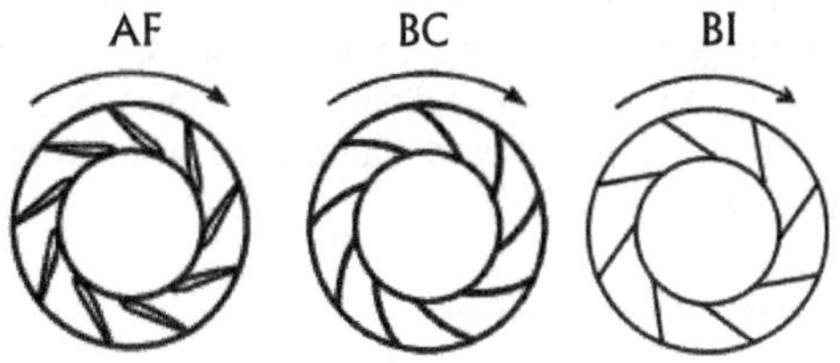

Housing Design

Cylindrical tube similar to vane axial fan, except clearance to impeller is not as close. Air discharges radially from impeller and turns 90° to flow through guide vanes. Variations include cylindrical and square housings with or without guide vanes. Square housings can include side discharge.

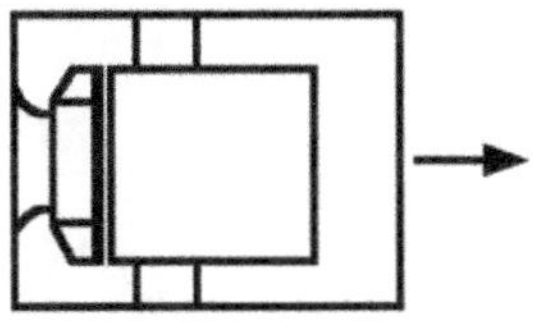

Performance Characteristics

Performance similar to backward inclined fan, except capacity and pressure are lower. Lower efficiency than backward inclined fan because air turns 90°.

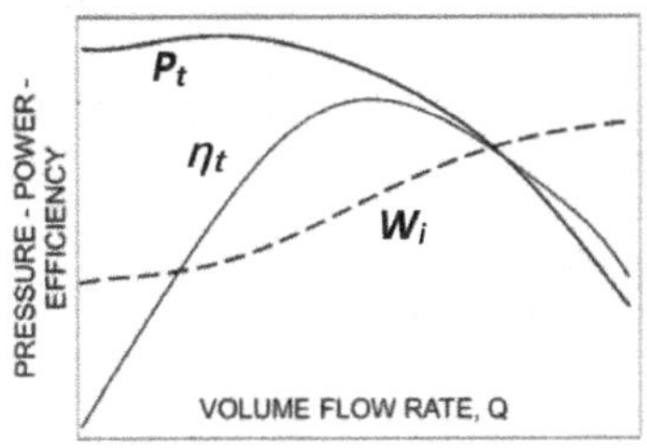

Applications

Ducted HVAC applications with air discharging in axial direction, such as low to medium pressure return air systems in HVAC applications.

Table 6-1 Types of Fans *(Continued)*

(Adapted from ASHRAE 2017, Table 1)

Power Roof Ventilator—Centrifugal

Impeller Design

Single inlet centrifugal impellers can be airfoil, backward curved, or backward inclined.

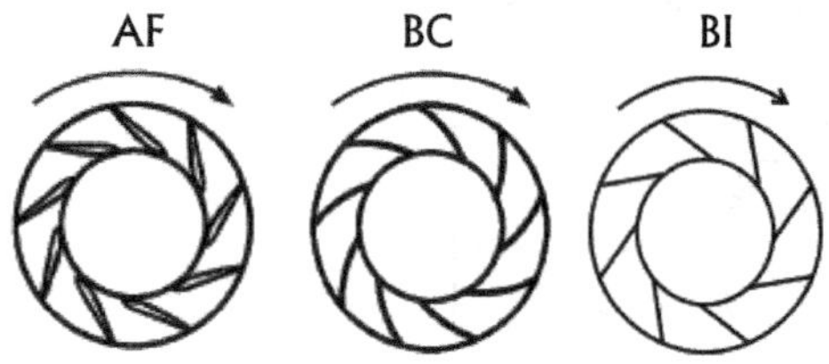

Housing Design

Weather-protected housing with means of mounting to a building opening. Usually does not include configuration to recover velocity pressure component. Radial discharge from impeller can be directed either toward building or away from building.

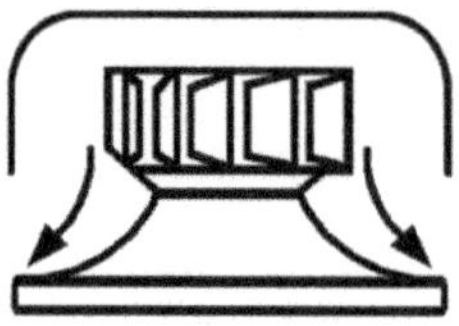

Performance Characteristics

Less efficient than scroll-type housed fan. Centrifugal units are slightly quieter than axial units.

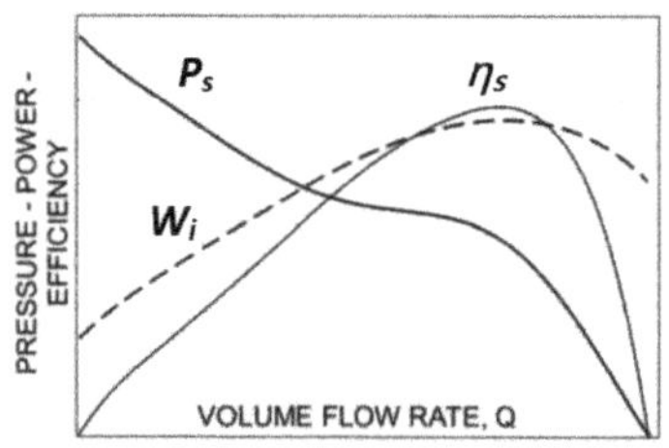

Applications

Low pressure ducted exhaust systems, such as general factory, kitchen, warehouse, and some commercial installations.

Power Roof Ventilator—Axial

Impeller Design

Small number of blades (2-6), similar to an axial propeller fan.

Housing Design

Similar housing to an axial propeller fan with the addition of weather protection and means of mounting to a building opening. Can supply or exhaust air from a building and housing can direct air either toward building or away from building.

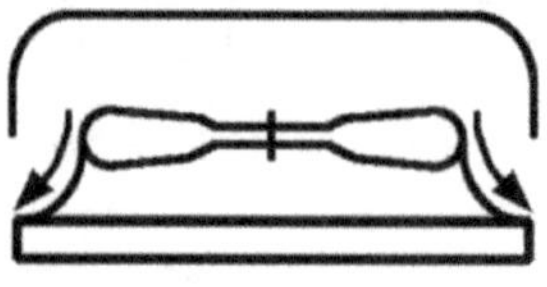

Performance Characteristics

Usually installed without ductwork; therefore, operates at very low pressure and high volume.

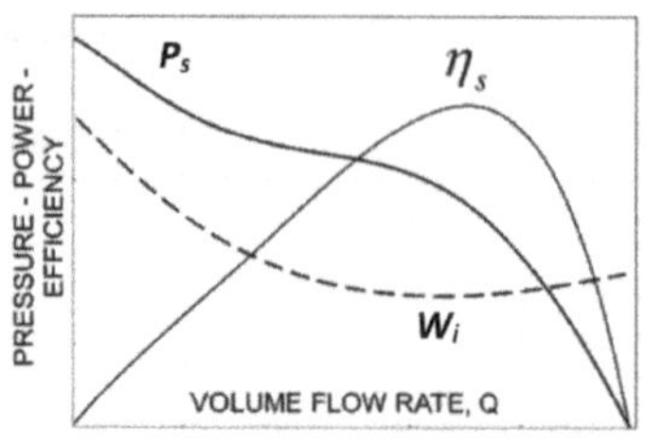

Applications

Low pressure exhaust or supply systems, such as general factory, kitchen, warehouse, and some commercial installations.

Table 6-1 Types of Fans *(Continued)*
(Adapted from ASHRAE 2017, Table 1)

<table>
<tr><th colspan="2">Induced Flow Exhaust</th></tr>
<tr><th>Impeller Design</th><th>Housing Design</th></tr>
<tr><td>Centrifugal backward inclined or mixed flow.
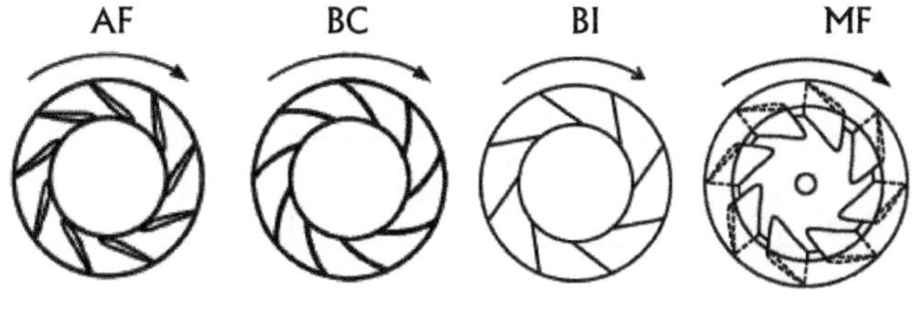
</td><td>Can be scroll or tubular inline with converging cone or nozzle to increase nozzle velocity which induces additional airflow through outlet.
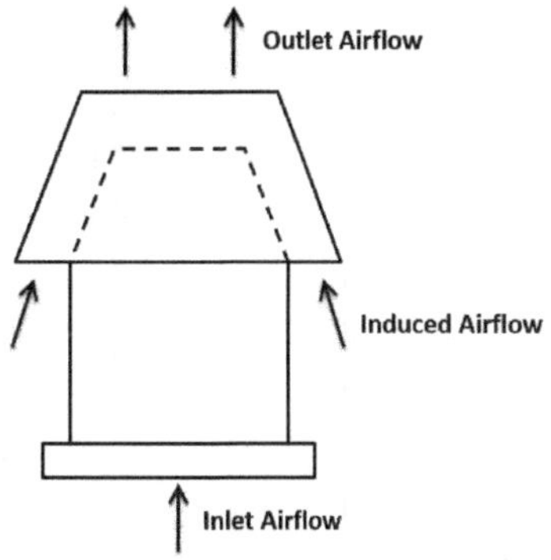
</td></tr>
<tr><th>Performance Characteristics</th><th>Applications</th></tr>
<tr><td>Performance similar to BI and MF but with reduced efficiency due to high velocity needed for induction. Outlet airflow greater than inlet airflow.
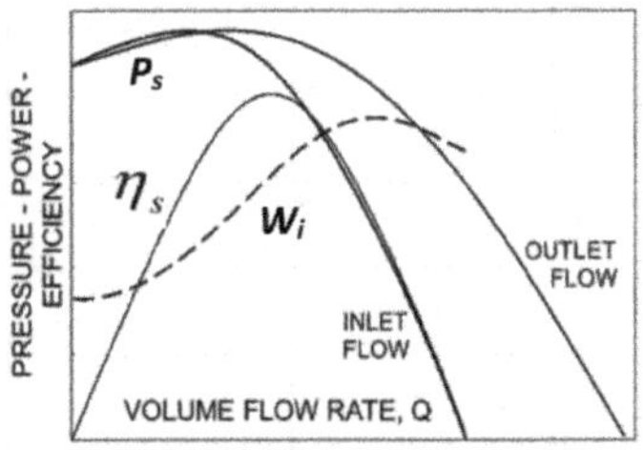
</td><td>Typically used for laboratory or hazardous chemical exhaust where dilution is required and tall stacks are not desired.</td></tr>
</table>

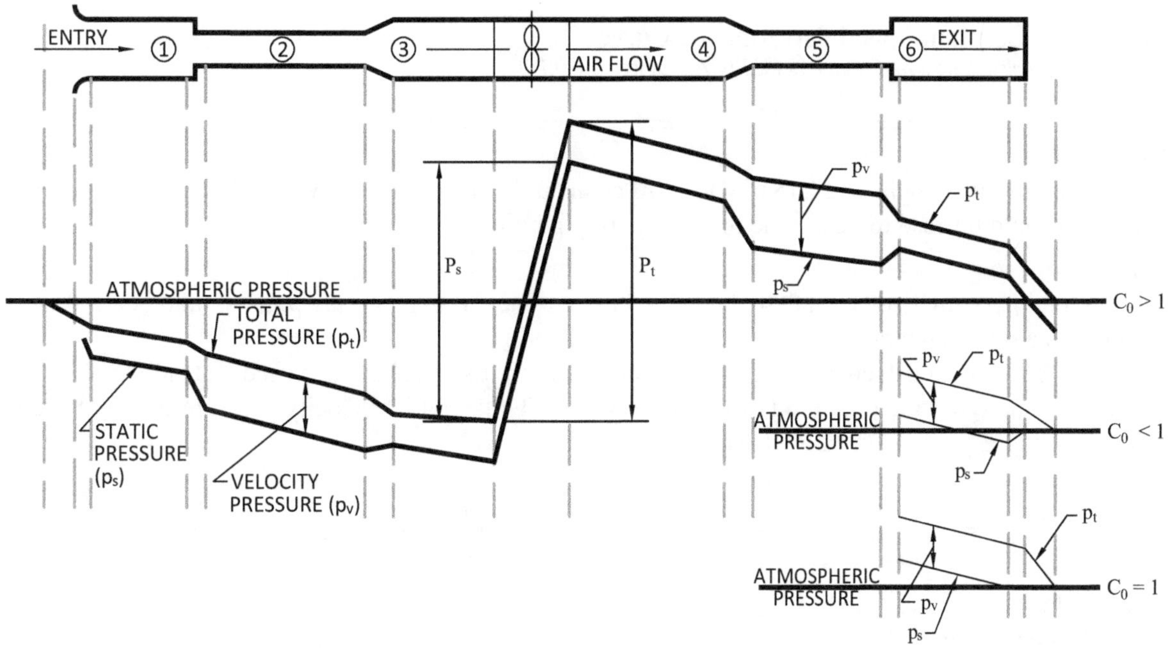

Figure 6-1 Pressure Changes During Flow in Ducts
(Adapted from ASHRAE 2017, Figure 7)

The difference between system total pressure and system fan static pressure can easily be seen in Figure 6-2. *System total pressure* corresponds to the system resistance as calculated by the design procedures in this design guide and should be used to determine the fan design operating point for applications with ducted fan inlets and outlets or when air-handler filters and coils are present. *Fan static pressure* should be used instead for applications without ducted outlets (as stated in Table 6-1).

Example 6-1 covers how to calculate the total pressure rise that should be used to determine the fan operating point.

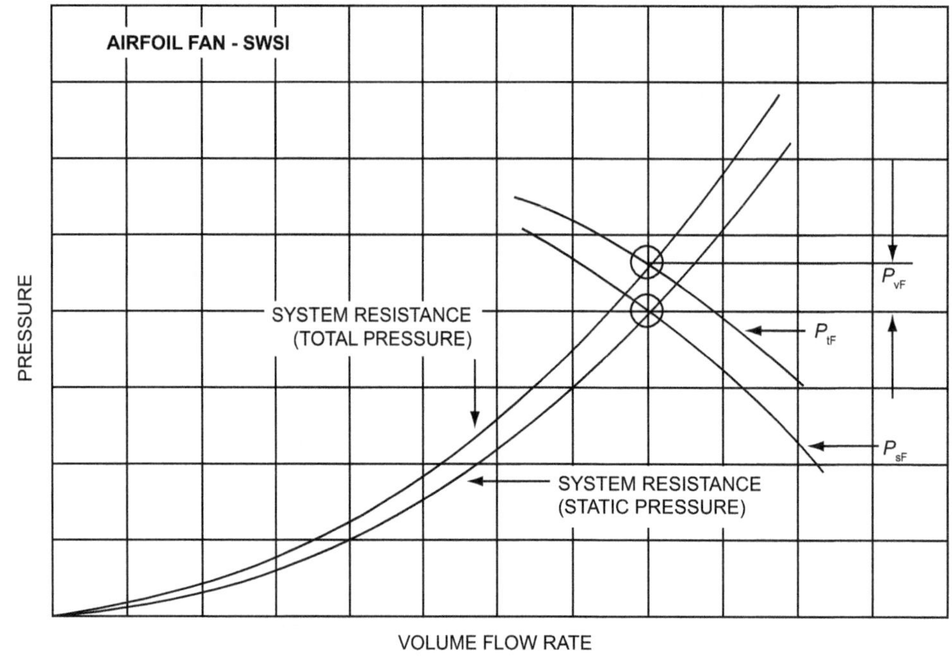

Figure 6-2 Constant-Speed Performance Curve
(Reprinted with permission from AMCA 200–95 (R2011), Air Systems (2011a))

Example 6-1.
For the single-inlet single-width (SWSI) airfoil centrifugal fan curves shown in Figure 6-2, what pressure rise should be used to determine the fan operating point at 8000 cfm?

Solution.
System total pressure rise P_t at 8000 cfm is 2.25 in. of water. This operating point should be used when the fan has a ducted outlet.

For a fan without a ducted outlet, the fan static pressure rise P_s should be used instead. As shown in Figure 6-2, the fan static pressure is 2.02 in. of water. The fan outlet velocity pressure is calculated as follows (using Equation 1-5a from Chapter 1):

$$p_v = \rho\left(\frac{V_o}{1097}\right)^2 = \rho\left(\frac{Q}{1097\,A_o}\right)^2 = 0.075\left[\frac{8000}{(1097)(4.19)}\right]^2 = 0.23 \text{ in. of water}$$

where velocity V is calculated by Equation 1-3 from Chapter 1 ($V = Q/A_d$).

In this case, fan total pressure is calculated as follows using Equation 6-1:

$$P_t = P_s + p_{v,\,o} = 2.02 + 0.23 = 2.23 \text{ in. of water}$$

FAN RATINGS

Most fans are rated in terms of fan static pressure and flow. However, fans having a high discharge velocity such as vane axial fans are often rated in terms of total pressure. Be aware of these different methods of rating and be certain whether fan static pressure or fan total pressure was used by a fan manufacturer.

The performance of fans is usually published in the form of a multi-rating table. A typical multi-rating table is illustrated in Figure 6-3.

Figure 6-4 shows constant-speed characteristic curves superimposed on a section of the multi-rating table for the same fan. A study of this figure can assist in the understanding of the relationship between fan curves and multi-rating tables.

DUCT SYSTEM CHARACTERISTICS

The pressure rise across the fan must be sufficient to overcome the total pressure drop of the air-handling system. Depending on the system configuration, the pressure drop is a function of the supply duct static pressure set point; outdoor air economizer operation; duct leakage; and the pressure drops across duct and duct-like elements (e.g., dampers, fittings), coils, and filters that are connected to the fan inlet and outlet. It is generally recognized that duct and duct-like pressure drops increase approximately as the square of the flow through them. However, pressure drops across coils and filters behave differently; the pressure drop versus flow relationship is less parabolic and more linear in some cases, such as with wet coils or high-efficiency filters (Sherman and Wray 2010).

IMPELLER DIAMETER	36-1/2 INCHES
OUTLET AREA	7.65 SQ FT
TIP SPEED IN FPM	9.56 X RPM
MAXIMUM BHP	18.3 X (RPM/1000)3

Volume	Outlet	1/4 in. wg		3/8 in. wg		1/2 in. wg		5/8 in. wg		3/4 in. wg		7/8 in. wg		1 in. wg		1 1/4 in. wg		1 1/2 in. wg	
CFM	Vel. (fpm)	rpm	bhp	rpm	bhp	rpm	bhp	rpm	bhp	rpm	bhp	rpm	bhp	rpm	bhp	rpm	bhp	rpm	bhp
3825	500	222	0.195																
4590	600	236	0.233	270	0.334														
5355	700	253	0.292	284	0.400	313	0.519												
6120	800	272	0.365	300	0.483	327	0.608	352	0.743										
6885	900	292	0.450	317	0.579	343	0.716	366	0.856	389	1.01	411	1.17						
7650	1000	314	0.560	337	0.695	360	0.840	382	0.992	403	1.15	424	1.31	443	1.48				
8415	1100	338	0.682	358	0.832	378	0.981	399	1.144	419	1.31	438	1.48	458	1.60	494	2.04		
9180	1200	361	0.826	379	0.988	398	1.149	417	1.314	436	1.49	455	1.68	472	1.86	507	2.25	540	2.67
9945	1300	385	0.989	402	1.163	419	1.340	437	1.514	454	1.69	472	1.89	489	2.09	522	2.49	554	2.92
10710	1400	409	1.175	425	1.360	441	1.553	457	1.741	473	1.93	489	2.12	506	2.34	538	2.76	568	3.20
11475	1500	434	1.387	449	1.587	464	1.780	479	1.993	494	2.19	509	2.40	524	2.61	555	3.06	584	3.52
12240	1600	458	1.626	473	1.837	488	2.048	501	2.269	515	2.49	529	2.70	543	2.92	572	3.39	600	3.87
13005	1700	483	1.895	498	2.115	511	2.346	525	2.570	537	2.80	550	3.03	564	3.26	590	3.73	617	4.24
13770	1800	508	2.191	522	2.424	535	2.665	538	2.901	560	3.15	572	3.40	585	3.64	610	4.12	635	4.63
14535	1900			547	2.767	559	3.017	571	3.275	584	3.52	595	3.78	606	4.04	630	4.55	654	5.07
15300	2000			571	3.144	584	3.403	595	3.672	607	3.93	618	4.21	629	4.48	651	5.02	674	5.56
16830	2200			621	4.003	633	4.289	644	4.577	654	4.87	665	5.16	675	5.46	695	6.06	715	6.65
18360	2400					682	5.335	693	5.632	703	5.96	712	6.28	721	6.61	741	7.24	759	7.90
19890	2600							742	6.885	752	7.22	761	7.56	769	7.91	788	8.60	805	9.30
21420	2800							791	8.308	801	8.67	810	9.03	818	9.40	834	10.15	852	10.88
22950	3000									850	10.32	859	10.71	867	11.09	883	11.89	898	12.70
24480	3200											908	12.50	916	13.01	932	13.84	946	14.70
26010	3400													965	15.16	981	16.03	995	16.92
27540	3600													1015	17.52	1030	18.47	1044	19.39
29070	3800															1072	21.16	1093	22.13
30600	4000															1129	24.11	1142	25.16

TYPICAL MULTISPEED RATING TABLE FOR A SINGLE WIDTH, SINGLE INLET CENTRIFUGAL FAN

Note: FSP = P_s

Figure 6-3 Typical Centrifugal Fan Performance Table

(Reprinted with permission from AMCA 201–02 (R2011), Fans and Systems *(2011b))*

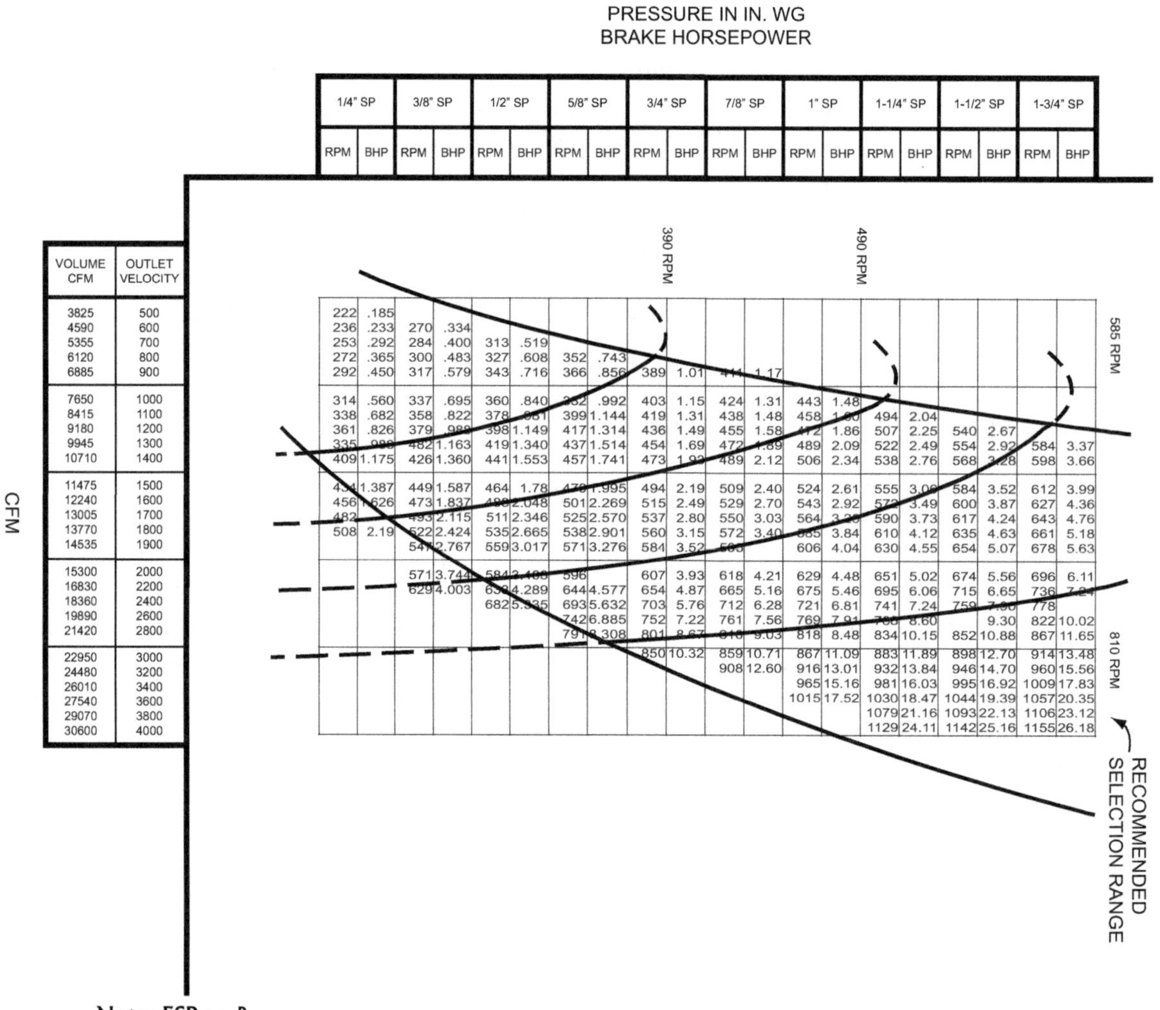

Note: FSP = P_s

Figure 6-4 Typical Fan Performance Table Showing Relationship to a Family of Constant-Speed Performance Curves
(Reprinted with permission from AMCA 201-02 (R2011), Fans and Systems *(2011b))*

For each combination of duct static pressure set point, economizer damper positions, leakage, and element pressure drops, it is possible to plot the associated fan pressure rise over a range of fan airflows for a given system operating condition. The overall relationship between pressure drop and flow defines what is commonly called a *system curve* as shown in Figure 6-5. When duct system characteristics change, such as when the duct static pressure set point is varied, a family of system curves can result.

The system-fan curve intersections that result when a system curve is plotted along with fan performance curves (e.g., power or speed as a function of pressure rise and flow) on a pressure versus flow map define a locus of unique fan operating points. Each of these points has an associated fan efficiency, power, and speed. The design operating point corresponds to the total pressure drop (resistance to flow) through the critical path or design leg of a duct system at the design airflow rate.

For illustrative purposes only, assume a duct system has no coils or filters and no leakage or air economizer and that the system curve is a simple quadratic function that passes through the origin. In this case, if the flow rate changes, the resulting total pressure varies as shown by Equation 6-2.

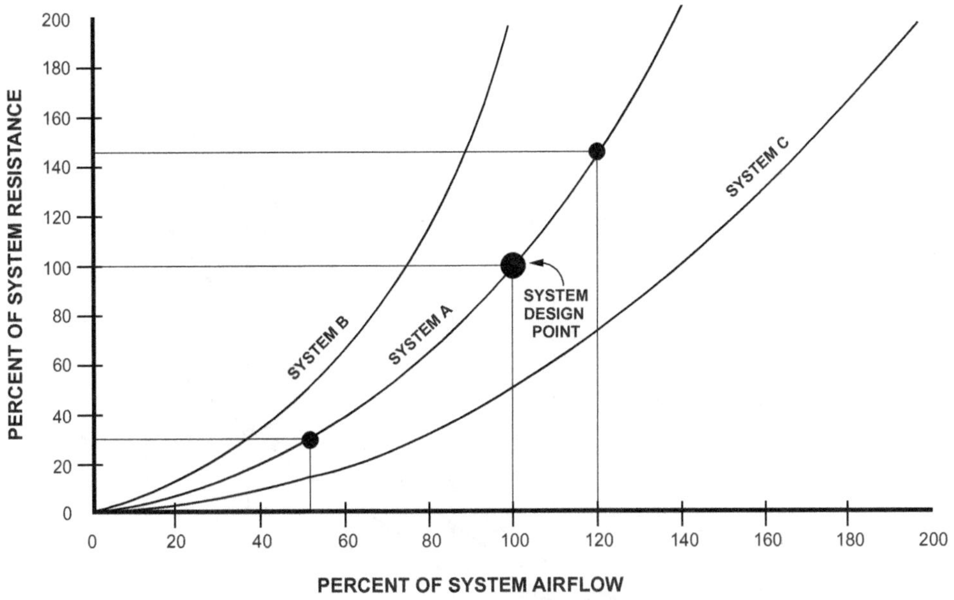

Figure 6-5 Normalized Duct System Curves
(Reprinted with permission from AMCA 201–02 (R2011), Fans and Systems *(2011b))*

This equation permits plotting a system's pressure characteristic curve from a calculated design condition. The fixed system must operate at some point on this system curve as the volume flow rate changes.

$$\left(\frac{\Delta p_2}{\Delta p_1}\right) = \left(\frac{Q_2}{Q_1}\right)^2 \qquad (6\text{-}2)$$

The characteristic curve of such a fixed system plots as a parabola in accordance with Equation 6-2. Typical plots of the resistance to flow versus volume flow rate for three different and arbitrary fixed systems (A, B, and C) are illustrated in Figure 6-5. For these systems, an increase or decrease in system resistance results from an increase or decrease in the volume flow rate along the given system curve only. For example, referring to duct system A in Figure 6-5 and assuming a system design point at 100% volume flow and 100% resistance, it can be seen that if the volume flow rate is increased to 120% of design volume, the system resistance increases to 144% of the design resistance in accordance with the system equation. A decrease in volume flow to 50% of design volume flow results in a decrease to 25% of the design resistance. Note that on a percentage basis, the same relationship also holds for duct systems B and C.

Example system total pressure loss curves are shown in Figure 6-6. Point A is the calculated design point for a system: 10,000 cfm at 3 in. of water. If these values are substituted into Equation 6-2 for Δp_1 and Q_1, other points of the system's characteristic curve can be determined. For 6000 cfm, the pressure loss is 1.08 in. of water calculated as follows (point D):

$$\Delta p_2 = 3\left(\frac{6000}{10{,}000}\right)^2 = 1.08 \text{ in. of water}$$

If a change is made within the system such that the total pressure at the design flow rate is increased, the system no longer operates on the previous system curve and a new curve is defined.

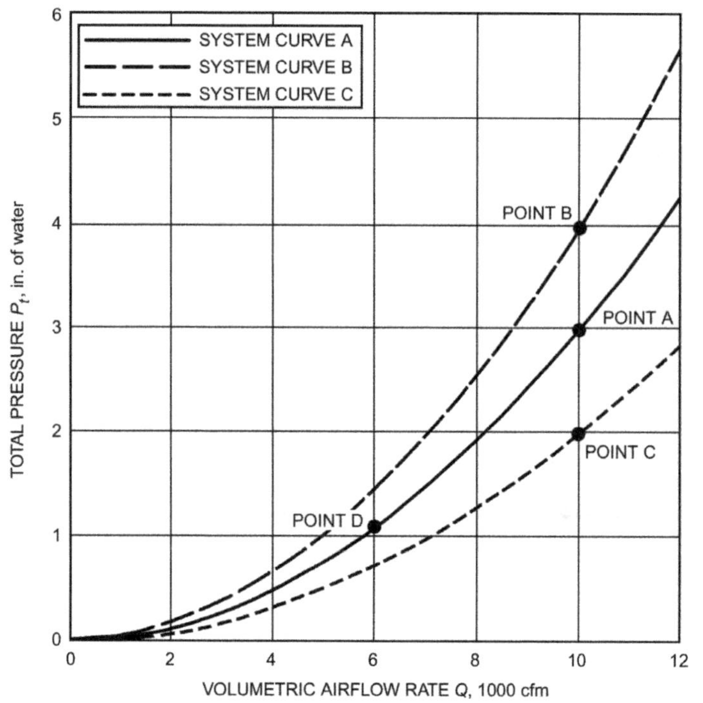

Figure 6-6 Example System Total Pressure Loss (Δp) Curves
(Reprinted from ASHRAE 2016b, Figure 10)

For example, in Figure 6-6, fittings added to the system's design leg or critical path increase the total pressure of the system at 10,000 cfm. If the total pressure is increased by 1.0 in. of water, the system total pressure drop at this point is now 4.0 in. of water, as shown for point B on system curve B. If instead the system pressure drop is decreased by changing the fittings in the design leg, the resulting system total pressure decreases, and system curve C is formed. For a 1.0 in. of water reduction at 10,000 cfm, the point of operation is 2.0 in. of water, as shown by point C.

In an actual operating duct system the three system curves in Figure 6-6 can represent three system characteristic curves caused by three different positions of an outlet damper or inlet guide vanes. Curve C is the most open position and curve B is the most closed. Figure 6-6, however, does not describe performance of a constant-pressure variable-air-volume (VAV) system. The system performance as implied by Equation 6-2 and Figures 6-5 and 6-6 applies only to simple systems without coils and filters and without duct static pressure control (i.e., return and exhaust systems). For constant-volume or VAV systems that include coils and filters and that may also control pressures at some points, the system resistance curves can deviate widely from Equation 6-2, even though many of the system elements may be described by this equation. A Lawrence Berkley National Laboratory (LBNL) report (Sherman and Wray 2010) provides more details about these types of systems, including the effects of system leakage. Also refer to Chapter 21, "Fans," of *ASHRAE Handbook—HVAC Systems and Equipment* (2016b) for further information.

Fan System Effects

Depending on the fan type and expected applications, fans for use with ducted systems are tested in a laboratory with an open inlet and a straight section of duct connected to the fan outlet, with an open outlet and a straight section of duct connected to the fan inlet, or with a straight section of duct connected to the fan inlet and one connected to the fan outlet, such that the fan has uni-

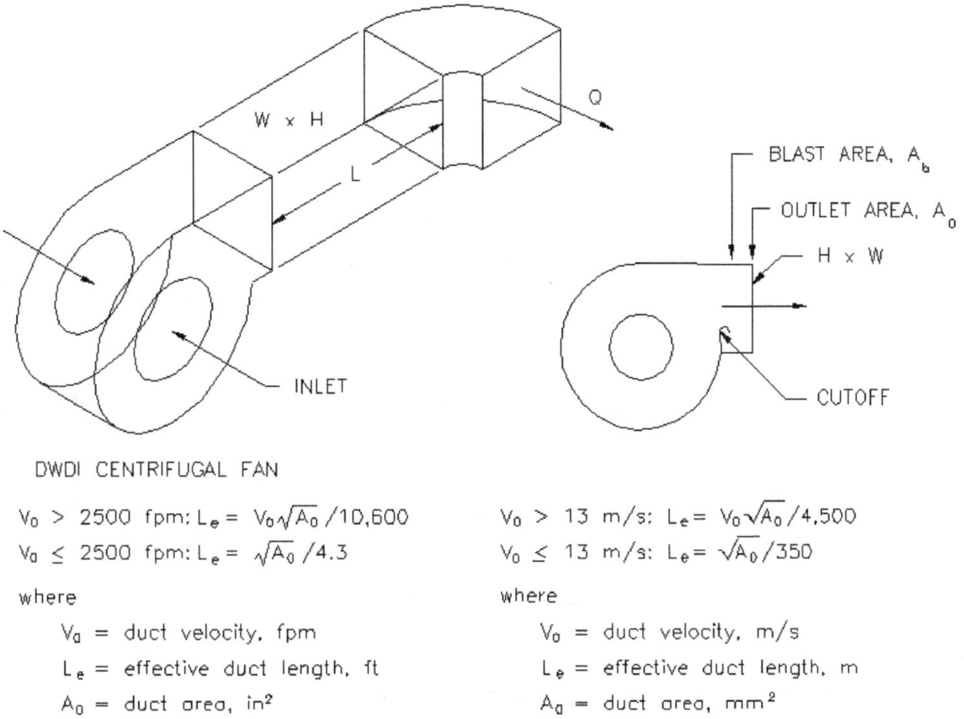

Figure 6-7 Illustration of Elbow Position at Fan Outlet that Can Cause a System Effect
(Reprinted from ASHRAE 2016a, fitting SR7-10)

form inlet and outlet flow conditions. If these conditions are not present in an actual installation because of disruptions, the fan's installed performance will not match the rated conditions.

Fan inlet and outlet disruptions are accounted for by including an additional resistance in calculating the duct system curve. These additional losses are categorized in the ASHRAE Duct Fitting Database (DFDB; ASHRAE 2016a) as "Fan & System Interactions" (SR7 and ER7 series). An example of an outlet disturbance causing a system effect is shown in Figure 6-7, where the fan has an elbow at its outlet. If the duct length between the fan outlet and the elbow is adequate, the total pressure loss due to a fan system effect is zero. However, if the elbow connects directly to the fan, there can be a significant system effect. There should be at least five hydraulic diameters between the disruption and the fan inlet or fan outlet to avoid a system effect. See AMCA 201 (2011b) for more information on system effects.

Fan Laws

Effect of Change in Fan Speed

According to Chapter 21, "Fans," of *ASHRAE Handbook—HVAC Systems and Equipment* (2016b), the fan laws relate performance variables such as pressure rise and power for geometrically, kinematically, and dynamically similar fans. It can be shown that complete dynamic similarity is attained when both the dimensionless fan flow coefficient Q/ND^3 and Reynolds numbers $Re = \rho ND^2/\mu$ are equal. The fan Reynolds number accounts for inertia and viscous effects. In practice, viscous effects are relatively unimportant for applications involving large fans, such that dynamic similarity is ensured when the fan flow coefficients of any two geometrically/kinematically similar fans are equal. This concept gives rise to the simplified fan laws as expressed in Equations 6-3 through 6-5. An increase or decrease in fan speed alters the volume flow rate through the

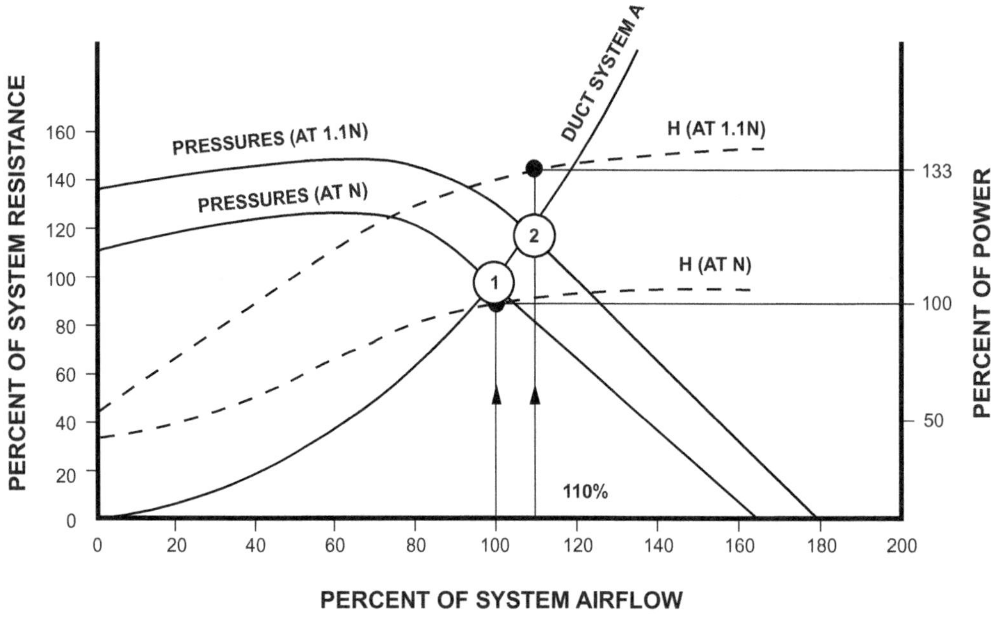

Figure 6-8 Effect of Fan Speed Change
(Reprinted with permission from AMCA 201–02 (R2011), Fans and Systems *(2011b))*

fan. For the same size fan ($D_2 = D_1$) and when air density does not vary ($\rho_2 = \rho_1$) the fan laws are as follows:

$$Q_2 = Q_1\left(\frac{N_2}{N_1}\right) \tag{6-3}$$

$$P_2 = P_1\left(\frac{N_2}{N_1}\right)^2 \tag{6-4}$$

$$W_{i2} = W_{i1}\left(\frac{N_2}{N_1}\right)^3 \tag{6-5}$$

It is important to recognize that the fan laws apply only to the fan and not to drive components such as belts, motors, or variable-frequency drives (VFDs). Variations in drive component efficiencies, especially at low loads, can impact the power required to operate the fan as the speed changes and are not accounted for in Equation 6-5.

Figure 6-8 illustrates the increase in flow rate when the speed of the fan is increased 10% to point 2. The 10% increase in flow rate results in a large power penalty. According to the fan laws, the fan shaft input power increase is 33%. If the connected drive system is not capable of a 33% increase in power, then insufficient airflow will occur.

Effect of Change in Density

The resistance of a duct system is dependent on the density of air flowing through the system. An air density of 0.075 lb_m/ft^3 is standard in the fan industry. Figure 6-9 illustrates the effect on fan performance of a density variation from standard air. For the same size fan, airflow, and speed, the

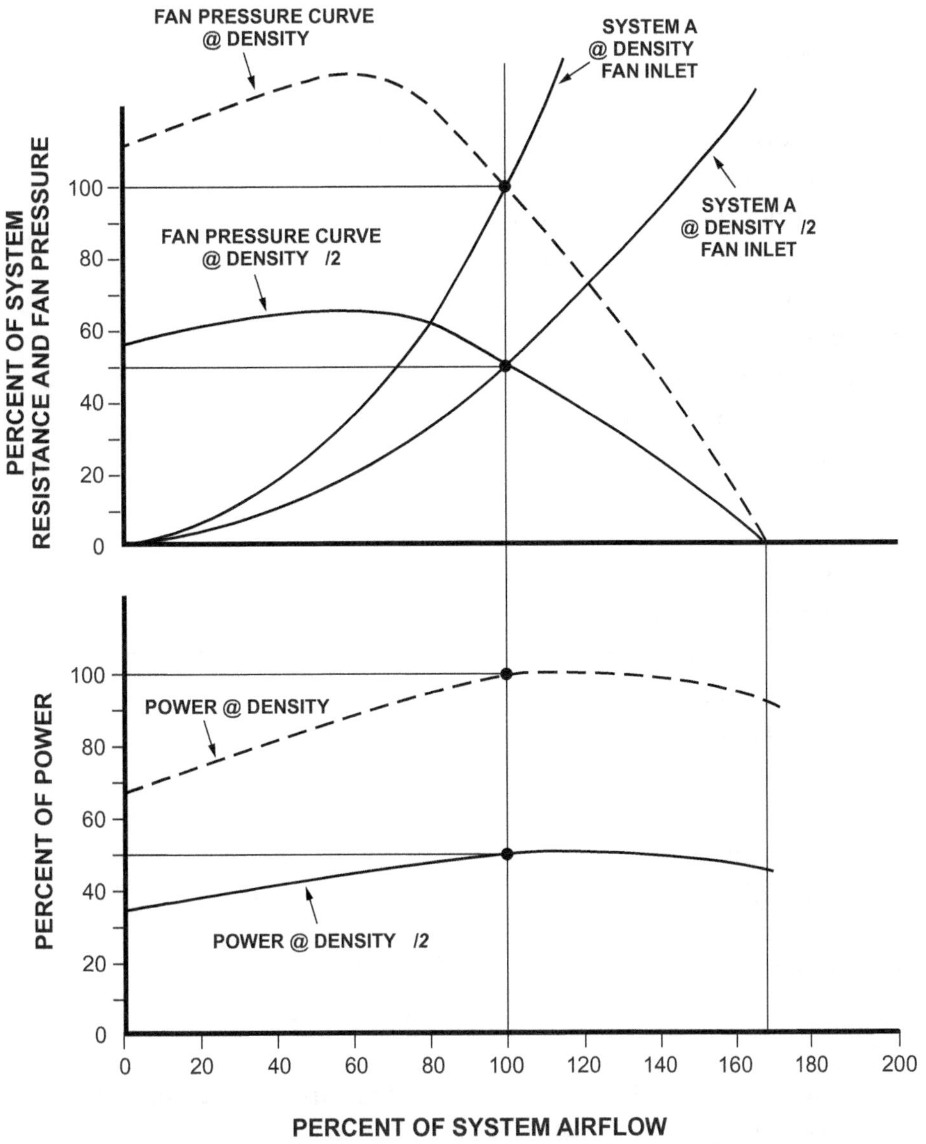

Figure 6-9 Density Effect
(Reprinted with permission from AMCA 201-02 (R2011), Fans and Systems *(2011b))*

pressure and power vary directly as the ratio of air density at the fan inlet to standard air density (Equations 6-6 and 6-7). The density ratio must always be taken into account when selecting fans from manufacturer's tables or curves. For the same size fan ($D_2 = D_1$) and airflow ($Q_2 = Q_1$) and when speed does not change ($N_2 = N_1$), the simplified fans laws are as follows:

$$P_2 = P_1\left(\frac{\rho_2}{\rho_1}\right) \tag{6-6}$$

$$W_{i2} = W_{i1}\left(\frac{\rho_2}{\rho_1}\right) \tag{6-7}$$

NOMENCLATURE

A	=	fan outlet area, ft^2
N	=	fan speed, rpm
P	=	fan pressure rise (static or total), in. of water
p_s	=	duct section fan static pressure, in. of water
P_s	=	fan static pressure, in. of water
p_t	=	duct section total pressure, in. of water
P_t	=	fan total pressure, in. of water
p_v	=	velocity pressure, in. of water
Q	=	fan airflow, cfm
V	=	fan outlet velocity, fpm
W_i	=	fan shaft input power, bhp

Symbols

Δp	=	system total pressure loss, in. of water
μ	=	viscosity, $lb_m/ft{\cdot}min$
ρ	=	fan inlet air density, lb_m/ft^3
η_s	=	fan static efficiency, %
η_t	=	fan total efficiency, %

Subscripts

1	=	known variable
2	=	calculated variable
i	=	input
o	=	fan outlet

REFERENCES

AMCA. 2011a. AMCA 200-95 (R2011), *Air systems*. Arlington Heights, IL: Air Movement and Control Association International.

AMCA. 2011b. AMCA 201-02 (R2011), *Fans and systems*. Arlington Heights, IL: Air Movement and Control Association International.

AMCA. 2016. ANSI/AMCA Standard 210-16 (also ANSI/ASHRAE Standard 51-16), *Laboratory methods of testing fans for certified aerodynamic performance rating*. Arlington Heights, IL: Air Movement and Control Association International.

ASHRAE. 2016a. ASHRAE Duct Fitting Database, ver. 6.00.05 (online). Peachtree Corners, GA: ASHRAE. www.ashrae.org/technical-resources/bookstore/duct-fitting-database.

ASHRAE. 2016b. Chapter 21, Fans. In *ASHRAE handbook—HVAC systems and equipment*. Peachtree Corners, GA: ASHRAE.

ASHRAE. 2017. Chapter 21, Duct design. In *ASHRAE handbook—Fundamentals*. Peachtree Corners, GA: ASHRAE.

Sherman, M.H., and C.P. Wray. 2010. *Parametric system curves: Correlations between fan pressure rise and flow for large commercial buildings*. LBNL Report Number 3542E.Berkeley, CA: Lawrence Berkeley National Laboratory.

BIBLIOGRAPHY

AMCA. 2011. AMCA 203-90 (R2011), *Field performance measurement of fan systems*. Arlington Heights, IL: Air Movement and Control Association International.

AMCA. 2018. ANSI/AMCA Standard 208-18, *Calculation of the fan energy index*. Arlington Heights, IL: Air Movement and Control Association International.

ISO. 2010. ISO Standard 12759:2010, *Fans—Efficiency classification for fans*. Geneva: International Organization for Standardization.

7

Duct System Materials and Construction

OVERVIEW

Duct systems have a primary function of conveying air as well as a broad range of secondary functions and conditions that typically dictate the specifics of materials to be used for their construction and their method of fabrication. Many materials can be used to meet the Owner's Project Requirements (OPR) for a duct system.

The latest edition of the Sheet Metal and Air Conditioning Contractors' National Association (SMACNA) *Round Industrial Duct Construction Standards* (2013) is an excellent reference for various metals used to construct ductwork. Table 7-1 summarizes common materials used in duct construction, Table 7-2 lists typical material applications, and Table 7-3 lists various common materials used for thermal and/or acoustical considerations and their applications.

After the dimensional requirements of the ductwork system are determined, it is incumbent on the design engineer to convey the design intentions and requirements to the installing contractor and ductwork fabricator. Unless the engineer has overriding concerns, the ductwork fabrication and installation are usually delegated with reference to an appropriate SMACNA duct construction standard.

Functional requirements limiting the allowable deflections, material stresses, objectionable self-generated noise and vibration, air leakage, dimensional stability, and other construction considerations are the underlying basis for the SMACNA duct construction standards, which conveniently break down the functional requirements into tables, giving the fabricator an array of choices in regard to the selected gauge, joining systems, and reinforcements used to satisfy the conditions of service.

Many duct specifications delegate design to one or more standards or references. Designers should familiarize themselves with the specifics and limitations of any reference they cite. Specifications are often inadequate for nonstandard applications, and multiple references may contradict each other. References can also conflict with the specifications and drawing notes and/or details. It is very important to specify which portion of a reference applies to a project or takes precedence, as the references usually cover a broad range of applications. It is also common to note that when a conflict does arise, the more stringent of those referenced should be used.

CONDITIONS OF SERVICE

Structural

Duct systems are generally expected to last the lifetime of the facility in which they are installed. Most duct construction references start with the goal of choosing a method of construc-

Table 7-1 Metal Ductwork

Material	Coating Composition	Maximum Temperature	Atmospheric Corrosion Resistance	ASTM Specification
Aluminized steel, Type 1	Approximately 91% aluminum, 9% silicon (coating weights from 0.13 to 0.60 oz./ft^2 total both sides)	900°F without discoloring, 1050°F max	Superior to galvanized steel	A463
Aluminum 3003-H14	n/a	400°F	Good	B209
Copper 110	n/a	Unknown	Excellent	B152
Galvalume® steel (Aluminized steel, Type 1)	Aluminum-zinc alloy (55% aluminum and 45% zinc)	1050°F	Good	A924
Galvanized steel, G60	Zinc (0.6 oz/ft^2 total both sides)	400°F	Good	A924/A653
Galvanized steel, G90	Zinc (0.9 oz/ft^2 total both sides)	400°F	Good	A924/A653
Galvannealed steel	Zinc A60	400°F	Good	A924/A653
Mill phosphatized galvanized steel	Zinc phosphate	400°F	Good	A924/A653
Stainless steel, Type 304	n/a	752°F	Excellent	A480/A480M
Stainless steel, Type 304-L	n/a	752°F	Excellent	A480/A480M
Stainless steel, Type 316	n/a	752°F	Excellent	A480/A480M
Stainless steel Type 316-L	n/a	752°F	Excellent	A480/A480M
Stainless steel, Type 430	n/a	752°F	Good to excellent	A480/A480M
Stainless steels Type 444	n/a	752°F	Excellent	A480/A480M
Cold rolled steel (CRS)	n/a	650°F	Rusts readily	A1008A/1008M and A568
Hot rolled steel (HRS)	n/a	650°F	Rusts readily	A36/A635/A1011

tion that will produce a duct system that will not structurally fail over the expected life of the system. An example is the latest edition of SMACNA's *HVAC Duct Construction Standards—Metal and Flexible* (2005). This comprehensive reference contains information on construction specifications for rectangular, round, and flat oval ductwork as well as casings and a variety of special applications. Chapter 11 of the manual includes functional criteria that establish the acceptable limits for burst pressure and collapse pressure, wall deflection, air leakage, and suspension from the building structure for ductwork and components that are not tabulated in the manual. Building codes, energy codes, and customer requirements may impose additional structural criteria.

Thermal

Extreme temperature conditions (1000°F or higher) dictate special materials and design considerations. Specialized systems such as kitchen grease exhaust systems call for a duct system to

Table 7-2 Typical Applications for Metal Ductwork

Material	Typical Applications
Steel, cold, or hot rolled	Appropriate for duct systems that are not used very often; will rust quickly. Not recommended for handling corrosive gases, vapors, or mists.
Steel, cold, or hot rolled, coated	Lasts longer than duct systems that are uncoated. System life is dependent on coating used and duct metal thickness. Must be cleaned before painting; all grease and oil must be removed with mineral spirits or other methods. Check with coating manufacturers for compatibility and maximum expected operating temperature.
Galvanized steel[1]	Successfully used for standard HVAC systems, including outdoor use and exposed duct systems. Zinc coatings offer a very good combination of galvanic and corrosion barrier protection.[2] G60 is standard for indoor applications. G90 gives additional protection for outdoor use. Not recommended for systems conveying corrosive aerosols or for chlorinated pool areas.
Galvanized steel, coated	Typically used for exposed duct systems in restaurants, schools, and other places. Coating galvanized steel that has been chromate treated may cause the paint to delaminate. Ductwork must be properly cleaned; consult coating manufacturers for proper pretreatment and coating systems. Refer to GalvInfoNotes 2.11 and 2.12 (Zinc 2017b, 2017c).[3]
Galvannealed steel, coated	Should be painted shortly after installation to prevent surface rust. Performance for coating adhesion is synergistically improved because of the excellent bond formed between the galvannealed surface and the coating. GalvInfoNote 1.3 (Zinc 2018)[4] explains how the hot-dip galvannealed manufacturing process differs from the hot-dip galvanized manufacturing process. It is extremely important that surfaces be clean and dry before painting. Surface dirt or rust must be removed with a stiff wire brush. Grease and oil must be removed with mineral spirits or detergent and water.
Polyvinyl chloride (PVC) precoated galvanized steel	Primary applications are buried ducts, ducts encased in underground tunnels, and under-slab ducts. The PVC coating needs to be touched up with a PVC touch-up paint after manufacture (and probably after installation) due to cut edges, the processes used in the manufacturing of ductwork, and scratches from shipping and handling. Due to the PVC coating, it is not possible to stitch weld or spot weld this material, and fittings are usually fabricated with rivets, button punches, and/or screws and sealed with a mastic. Do not use PVC-coated galvanized steel for corrosive or condensing-fume exhaust applications.
Powder coat galvanized steel	Powder coating involves using 100% dry resin that is electrostatically applied to the surface and then baked in an oven to fuse the coating. A variety of coating materials and colors are available for aesthetic and/or chemical resistance.
Aluminum[5]	Appropriate for swimming pool areas and rural, industrial, and marine atmospheres. Aluminum requires a 44% increase in thickness to match the deflection performance of galvanized steel. See the galvanized steel gauge equivalents for aluminum in Table 2-50 of SMACNA's HVAC Duct Construction Standards—Metal and Flexible (2005).
Galvalume® and aluminized Type 1	Used for heat-resisting applications such as furnace exhaust and also for uses where corrosion resistance is involved. Use instead of galvanized steel for temperatures above 400°F. This process ensures a tight metallurgical bond between the steel sheet and its aluminum coating, producing a material with a unique combination of properties possessed alone by neither steel nor aluminum.
Stainless steel	Stainless steel contains sufficient chromium to form a film of chromium oxide that prevents surface corrosion by blocking oxygen diffusion to the steel surface and blocking corrosion from spreading to the metal's internal structure. Types 304, 304-L, 316, and 316-L are commonly used for ducts in applications where concerns may include resistance to corrosion, heat, and abrasion. Stainless steels are available with different finishes for aesthetic reasons on exposed applications. A 2B finish is the most common. Types 316 and 316-L are more expensive than Types 304 and 304L and are generally used where specific chemical resistance applications are required. Type 316 has far superior corrosion resistance than Type 304.

[1] See GalvInfoNote 1.1, "Understanding Coating Weight Designations for Zinc-Based Coatings on Steel Sheet" (Zinc 2019), from https://www.galvinfo.com/galvinfonotes/. GalvInfo Center, a program of the International Zinc Association, was established in 1999 to serve current and potential users of steel sheet coated with zinc-containing coatings. Years to first rust for G90 coating outdoors average 3 to 7 for severe industrial, 15 to 20 for rural, 7 to 10 for Atlantic coast marine, 12 for suburban, and 10 for urban. Without condensation, high humidity, chemical exposure, or major temperature changes, the small amount of corrosion that occurs causes only discoloration in 20 to 30 years.
[2] See GalvInfoNote 2.1, "The Continuous Hot-Dip Coating Process for Steel Sheet Products" (Zinc 2017a).
[3] See GalvInfoNote 2.11, "Preparing Galvanize for Field Painting," and GalvInfoNote 2.12, "Pretreatments for Metallic-Coated Sheet" (Zinc 2017b, 2017c).
[4] See GalvInfoNote 1.3, "Galvanneal – Differences from Galvanize" (Zinc 2018).
[5] Aluminum has high corrosion resistance in rural, industrial, and marine atmospheres.

contain the fire hazard as well as prevent potential combustion of materials adjacent to the duct. Even common HVAC ducts have a thermal restraint to not exceed 120°F due to the sealants typically used in assembly and fabrication.

The latest editions of energy standards such as ANSI/ASHRAE/IES Standard 90.1, *Energy Standard for Buildings Except Low-Rise Residential Buildings* (ASHRAE 2019a), establish the thermal efficiency requirements for duct systems.

Under air-conditioning situations, engineering judgment must be used to prevent condensation from forming on the ductwork, especially that exposed in the conditioned space, from dripping onto the floor.

Acoustical

The reduction or exclusion of acoustic sound energy in a duct system can happen in more than one way. Rigid ducts and their associated mass reflect external noise but also tend to transmit internal noise for longer distances before it is dissipated or reflected. Insulation, laggings, and coatings can reduce the transmitted sound energy as well as breakout noise from a duct system.

Nonrigid ducts are sometimes referred to as *sound attenuating* ducts because they have low mass and therefore do not transmit acoustic sound energy well. Some acoustic sound energy is lost as breakout noise. This may be misleading, however, since it is possible that desired acoustical performance can be achieved if the acoustic sound energy breaks out in a location, such as above a ceiling, where it is not objectionable.

Conveyed Substances

Duct systems sometimes convey substances along with the air that may be abrasive, corrosive, combustible, flammable, or odorous or that may pose a life safety risk. These applications are outside of traditional HVAC comfort cooling and heating and are categorized as "industrial." Combustible applications may dictate nonferrous construction (nonsparking) or the inclusion of systems to wet the particulate. Flammable, odorous, and life-safety-risk applications can dictate material type, material thickness, and transverse connections and can require fully welded construction.

Design guidance on industrial duct systems can be found in references such as the SMACNA *Round Industrial Duct Construction Standards* (2013) and *Rectangular Industrial Duct Construction Standards* (2011) and the American Conference of Governmental Industrial Hygienists (ACGIH) *Industrial Ventilation: A Manual of Recommended Practice for Design* (2019). Information on corrosion resistance can be found in a number of sources, such as *Corrosion Resistant Materials Handbook* (De Renzo and Mellan 1985) or the Cole-Parmer Chemical Compatibility Database.[1]

Environmental

Some duct systems must be designed for external environmental conditions that may affect the durability or efficiency of the duct. For example, maritime environments and natatoriums have external conditions as the primary duct construction determinant instead of the conveyed air. Wind and seismic criteria also influence the selection of duct type and method of construction. The ASHRAE Handbooks are an excellent source of further information, particularly the HVAC Applications and HVAC Systems and Equipment volumes (2019c, 2016c).

Visual

Increasingly, duct systems have become part of the architectural design of a building. This greatly affects the selection of duct material, the shape, and the other construction choices. Owners,

1. www.coleparmer.com/chemical-resistance?searchterm=chemical+resistance

Table 7-3 Thermal and Acoustical Materials

Material	Composition	Maximum Temperature	Thicknesses	ASTM Specification*	Notes/Applications
Glass and Mineral Fiber					
Fiberglass duct liner	Glass fibers bonded with a thermosetting resin. Airstream surface protected with a coating.	250°F	0.5, 1, 1.5, and 2 in.	C1071	Internal liner typically used for rectangular HVAC ducts. Can be used in systems operating at velocities up to 6000 fpm.
Fiberglass duct wrap	Glass fibers bonded with a thermosetting resin. Exterior surface protected with a vapor-retarding facing.	250°F	1.5 to 4 in.	C1290	Used as an external insulation wrap for ducts of all shapes. Has the thermal benefits of liners but without the airstream contact of liners.
Fiberglass (unfaced)	Glass fibers bonded with a thermosetting resin.	350°F to 650°F	1 to 3 in.	C1139	Used for HVAC applications where the insulation is encapsulated on both the interior and exterior (double-wall ducts and panels) and not in the airstream.
Fiberglass duct board	High-density glass fiber board with airstream surface coating and exterior foil-scrim-kraft (FSK) facing.	250°F	1, 1.5, and 2 in.	C1071	Used for lining round ducts without the need for an internal metal shell. Also used for rectangular ducts and equipment where a more rigid insulation is desired.
Mineral fiber/ rock wool	Stone or silica heated until molten, then spun and formed into a flexible, fibrous mat.	up to 1000°F	Various	C1696	Lower water absorption. Heat resistant. Used for dual-walled kitchen grease exhaust ducts. When encapsulated with scrim-reinforced foil, it is used as a fire wrap.
Fiber Free					
Flexible elastomeric foam (FEF)	Flexible natural or synthetic rubber.	200°F to 250°F	1, 1.5 and 2 in.	C1534	Used where "fiber-free" insulation is desired. Generally considered a closed-cell insulation; a slightly more open-celled product is available for round applications.
Phenolic foam	Rigid thermoset phenolic foam core faced with aluminum foil on both sides.	185°F	0.875, 1.1875, and 1.75 in.	C518	Closed cell. Lower weight than other foams. Surface burning characteristics superior to flexible elastomeric or rigid polyiso. Thermal performance 50% greater than FEF.
Polyester	Hypoallergenic polyester material bonded with an FSK facing.	250°F	1, 1.25, 1.5, and 2 in.	C518	Used where "fiber-free" insulation is desired.
Rigid polyiso (PIR)	Rigid thermoset polyisocyanurate foam core faced with aluminum foil on both sides.	176°F	25/32 and 1.1875 in.	N/A	Closed cell, low weight, and fiber free.

* See the Bibliography section for citation information for these specifications.
Note: The information in this table was taken from manufacturers' data; users should verify the properties with the respective manufacturer.

architects, and designers should be aware that such systems go beyond the basic functional design addressed in most duct construction references. It is incumbent on the design team to select specific visual features and properly convey the design intent to the ductwork fabricator so that the overall project goals can be achieved.

REFERENCE STANDARDS

Duct design engineers should refer to the latest editions of SMACNA's *Rectangular Industrial Duct Construction Standards* (2011) or *Round Industrial Duct Construction Standards* (2013) for special construction requirements such as elevated temperatures, imposed loads (insulation, lagging, wind, snow, ice, particulate accumulation, and abrasive particulates in the airstream), corrosive gases, extended supports, and extreme environmental conditions. Brief summaries of the relevant standards are included here:

- *HVAC Duct Construction Standards—Metal and Flexible* (SMACNA 2005)
    - This standard covers the fabrication of rectangular, round, and flat oval ductwork and applies to most HVAC systems that are conveying clean air without any particulates, provided that
        - they are operating at a pressure ±10 in. of water (except aluminum duct, which is limited to ±2 in. of water),
        - temperatures are less than 120°F, and
        - the maximum support spacing is 10 ft for rectangular duct and 12 ft for round and flat oval duct.
    - For corrosive vapors, special consideration needs to be made for the composition of the duct material, the fabrication techniques for the duct and fittings, sealants (if any), and the transverse connections.
- *Round Industrial Duct Construction Standards* (SMACNA 2013)
    - This standard applies to both round spiral and round welded ducts, provided that
        - elevated temperatures do not exceed 1500°F;
        - the ducts are subject to wind, snow, and ice loading;
        - the supports are required to handle external loads on the ducts, such as lagging, insulation, and a maintenance person;
        - the pressures range between –30 and 50 in. of water;
        - extended support lengths are less than 30 ft; or
        - the ducts convey both corrosive and abrasive substances in the airstream.
- *Rectangular Industrial Duct Construction Standards* (SMACNA 2011)
    - This standard applies to rectangular duct only, provided that
        - design pressures range from –150 to 150 in. of water,
        - elevated temperatures do not exceed 1500°F,
        - extended support lengths are less than 30 ft, or
        - the ducts convey both corrosive and abrasive substances in the airstream.

Other references from SMACNA include *Phenolic Duct Construction Standards* (2015), *Fibrous Glass Duct Construction Standards* (2003), *Thermoplastic Duct (PVC) Construction Manual* (1995), *Thermoset FRP Duct Construction Manual* (2016), and *Kitchen Ventilation Systems and Food Service Equipment Fabrication and Installation Guidelines* (2001). See the SMACNA store (www.smacna.org/store) for a complete listing of their duct construction manuals.

OTHER CODES AND STANDARDS THAT MAY IMPACT DUCT MATERIALS AND CONSTRUCTION

The following codes and standards may also impact the selection of materials for duct construction:

- ANSI/ASHRAE/IES Standard 90.1, *Energy Standard for Buildings Except Low-Rise Residential Buildings* (ASHRAE 2019a)
- ANSI/ASHRAE Standard 62.1, *Ventilation for Acceptable Indoor Air Quality* (ASHRAE 2019b)
- *International Green Construction Code*® (*IgCC*; ICC 2018b)
- ANSI/ASHRAE/IES Standard 90.2, *Energy-Efficient Design of Low-Rise Residential Buildings* (ASHRAE 2018)
- *Seismic Restraint Manual—Guidelines for Mechanical Systems* (SMACNA 2008b)
- *Seismic Restraint Manual—Guidelines for Mechanical Systems—OSHPD Edition* (developed to meet the specific requirements of the California Office of Statewide Health Planning and Development for hospital and health-care facilities construction; SMACNA 2009)
- *IAQ Guidelines for Occupied Buildings Under Construction* (SMACNA 2008a)
- *International Building Code* (*IBC*; ICC 2018a)
- *International Mechanical Code* (*IMC*; ICC 2018c)
- *Industrial Ventilation: A Manual of Recommended Practice for Design* (ACGIH 2019)
- NFPA 91, *Standard for Exhaust Systems for Air Conveying of Vapors, Gases, Mists, and Particulate Solids* (2015)

DESIGN REQUIREMENTS

For typical HVAC systems providing comfort heating and cooling for people in occupied spaces, the temperature, system design pressure, and ductwork hangers are adequately covered by the latest edition of SMACNA's *HVAC Duct Construction Standards—Metal and Flexible* (2005). In addition to any insulation requirements, whether it is internally or externally lined, the maximum (positive or negative) operating pressure, design pressure, test pressure, and maximum amount of system air leakage need to be specified. The construction requirements of the ductwork should be based on the maximum static pressure that it will experience. At the fan, this is the positive static pressure at the outlet or the negative static pressure at the inlet. This is not the fan static pressure as defined in Chapter 6.

Hence, duct construction requirements should be guided by the following considerations:

- *Design Operating Static Pressure.* This is the maximum operating static pressure under system maximum design conditions. The duct construction class is based on the maximum operating pressure that the section of duct will see. Often, though, the duct construction class is based on this static pressure at the fan outlet and separately at the fan inlet to determine the construction of the ductwork.
- *System Air Leakage.* Refer to Chapter 19, "Duct Construction," in *ASHRAE Handbook—HVAC Systems and Equipment* (2016c) for system sealing and leakage testing requirements.

CLEANING

Ducts should be clean before use. Prior to installation, debris and liquids should be removed from the interior of the duct. During installation, open ends should be covered when not part of the active installation.

Many engineers and contractors have assumed that some elements from the highest level of cleanliness should be required on all jobs. The result has been higher costs with negligible or negative benefits to the owner. Ducts are typically transported to a job site within hours of fabrication. Water-soluble oils used in fabrication typically drain and evaporate if allowed, and duct sealants require time to dry and cure. But both of these processes are affected when duct is immediately capped for shipment. Therefore, it has become common to see internal moisture and white rust on ducts though the intent was to have them be clean throughout the shipment and job-site storage processes.

Designers should consider the correct level of cleanliness for their projects; further guidance is provided in SMACNA's *Duct Cleanliness for New Construction Guidelines* (2000), which outlines three appropriate levels of duct cleanliness.

DUCT SEALING

Refer to Chapter 19, "Duct Construction," in *ASHRAE Handbook—HVAC Systems and Equipment* (2016c) for system sealing requirements.

ANTIMICROBIAL DUCTS

Antimicrobial-treated ducts can be coated (before or after fabrication) with an antimicrobial agent that inhibits the growth of bacteria, mold, and fungi (including mildew). Textile, duct liner, and galvanized or stainless steel ducts can be coated if the service temperature of the antimicrobial compound is not exceeded. Textiles and liners can be made inherently antimicrobial by combining the antimicrobial chemistry into the manufacturing process or treating the finished fabric. Glass fiber duct liners and insulation are inorganic and inert, but the binding agents used in the manufacturing process are not and may act as a nutrient for mold growth but may still be coated with an antimicrobial compound. Many insulation products are being reformulated and obtaining certification for not breeding microbial growth per ASTM C1338 (2019).

All antimicrobial coatings or touch-up paint should be EPA-registered antimicrobial compounds for specific use in HVAC systems, be tested under ASTM E84 (2020), survive minimum and maximum service temperature limits, and comply with NFPA 90A and 90B (2018a, 2018b). Coatings should have flame spread/smoke developed ratings not exceeding 25/50.

Antimicrobial ductwork is considered a secondary protection as *the microbe must come in contact with the coating for it to be effective.* If microbes are airborne and do not contact the antimicrobial-coated surface, the antimicrobial protection will not be effective on those microbes; the outcome is the same if the duct surface has a coating of dust covering the antimicrobial layer.

ANSI/ASHRAE Standard 62.1, *Ventilation for Acceptable Indoor Air Quality* (ASHRAE 2019b), in section 5.4.1, "Resistance to Mold Growth," requires the material surfaces to be resistant to mold growth with a standardized test method (UL 181 [2013], ASTM C1338 [2019], or ASTM D3273 [2016a]); sheet metal and metal fasteners are exempt from this requirement.

COATING GALVANIZED DUCTS

Chemically treated galvanized steel can be coated or painted anytime with the proper cleaning and pretreatment. Pretreatment should be performed immediately before coating. Consult coating manufacturers for proper pretreatment and applicable coating systems. For a discussion of pretreatments for metallic-coated steel, see GalvInfoNote 2.12 (Zinc 2017c). ***Caution:*** Galvanized steel that has been chromate treated could be very difficult to coat. In some cases, aggressive pretreatments are necessary. Consult coating manufacturers for proper pretreatments and coating systems.

For any galvanized surface to be coated, it is extremely important that the surface be clean and dry. Surface dirt or rust must be removed with a stiff wire brush. Grease and oil must be removed with mineral spirits or detergent and water. All traces of soap should be removed by thorough rinsing. Coat only when the surface is completely dry. Sources of more detailed information on coating galvanized steel ductwork are ASTM D6386 (2016b), ASTM D7396 (2014), and GalvInfoNote 2.11 (Zinc 2017b).

PAINTING GALVANIZED DUCTWORK

A common pretreatment used to obtain good bonding between paint and galvanized steel ductwork is phosphate pretreatment. It can be applied to galvanized sheet by the steel manufacturer at the factory. This is often referred to as *mill phosphatized.* See GalvInfoNote 2.12 (Zinc 2017c) for additional information. G60 or G90 galvanized metal can be phosphatized.

Another type of galvanized treatment process is galvannealling, where galvanized steel undergoes an annealing process as described in GalvInfoNote 1.3 (Zinc 2018), which results in a hard, durable surface with 10% iron at the surface to provide for enhanced coating adhesion. Galvannealed metals are rated similar to standard galvanized metals but include an "A" designation such as A-40 or A-60. A-60 is the maximum coating that can be galvannealed.

Both phosphate pretreatment and galvannealed metals should be painted immediately after installation so the material does not start to corrode. Both have been generically called *paintgrip*, *paint grip*, *paintable galvanized*, and *pgrp*.

There are also high-performance water-based direct to metal (DTM) coating systems for standard G-60 or G-90 galvanized metal that can be used to paint galvanized metal without the phosphatizing or annealing processes.

Designers should ensure that what they want is properly called out in the specification (mill phosphatized, galvannealed, or standard galvanized painted with a DTM coating). See GalvInfoNote 1.3 (Zinc 2018) for relative corrosion rates of these pretreatment techniques.

WHITE RUST

Storage stain is a corrosion product that is typically white but can also take the form of a gray or black deposit on the surface of galvanized metal duct. Because the most common color of discoloration is white, storage stain is often called *white rust*. It can occur when sheets of galvanized steel that are in close contact (in a coil or stacked in lifts or bundles) get wet, either by water intrusion or by condensation from moist air trapped between the sheets. The discoloration is due to the corrosion products that form after the zinc reacts with moisture in the absence of free air circulation. Refer to GalvInfoNote 3.2 (Zinc 2017d) for more information on storage stain.

NONMETALLIC DUCTWORK MATERIALS

Fibrous Glass

Fibrous glass duct should conform to the latest SMACNA *Fibrous Glass Duct Construction Standards* (2003) or North American Insulation Manufacturers Association (NAIMA) *Fibrous Glass Duct Construction Standards* (2016). With this type of duct, duct board is made from glass fibers bonded with a thermosetting resin then covered on the exterior with a foil-scrim-kraft (FSK) facing. With densities of 3.8 lb/ft^3 and higher, these products are fairly rigid and can be used in systems operating in the range of ±2 in. of water and air velocities up to 2400 fpm. The maximum temperature inside the duct operating continuously is 250°F, and the maximum duct surface temperature should not exceed 150°F. Fibrous glass duct systems resist mold and fungi growth when tested in accordance with UL 181 (2013) and ASTM C1338 (2019) and should meet UL 181 Class I air duct requirements. Consult manufacturers for other limitations that may apply to specific products.

Phenolic Foam

Phenolic foam has thermal performance approximately 50% greater than fiberglass or elastomeric insulations of the same thickness. It is also greater than 90% closed cell and meets the 25/50 flame spread/smoke developed criteria. Phenolic duct systems are typically rectangular and can be provided in joint lengths of up to 13 ft. Ducts are fabricated from panels composed of a phenolic insulation core faced with aluminum foil on each side. SMACNA's *Phenolic Duct Construction Standards* (2015) contains instruction, tables, and details for basic phenolic duct fabrication and installation.

Flexible Ducts

A flexible duct is categorized by its listing as either an air duct or an air connector. Air ducts and air connectors may be metallic or nonmetallic, insulated or not insulated, and they should be listed and labeled as Class 0 or Class 1 per UL 181, *Standard for Factory-Made Air Ducts and Air Connectors* (2013). These ducts should be installed per the conditions of their listing and recommendations per the Air Duct Council's *Flexible Duct Performance & Installation Standards* (ADC 2010). The maximum allowed temperature inside these ducts is 250°F per their listing and NFPA 90A and 90B requirements (NFPA 2018a, 2018b). Flexible ducts have the following requirements:

- Air ducts are tested in accordance with UL 181 (2013) and must pass 16 separate tests. Air connectors are required to pass the same tests except the tests for flame penetration, puncture, and impact.
- Flexible air ducts and air connectors should be installed using the minimum length needed and without longitudinal compression. The most practical route between connection points should be selected while not overly stressing the material or attempting to remove all available stretch. Excess length should not be left for future building modifications. Flexible air ducts and air connectors should be properly supported per the manufacturer's installation instructions.
- Flexible air connectors are intended for limited use within an HVAC system. They are limited by their listing, the *International Mechanical Code®* (*IMC*; ICC 2018c), and NFPA 90A and 90B (2018a, 2018b) to maximum 14 ft installed length. The *Uniform Mechanical Code®* (IAPMO 2018) prohibits use of air connectors.
- For commercial duct systems, Chapter 21, "Duct Design," of *ASHRAE Handbook—Fundamentals* (2017a) recommends that flexible duct length be limited to a maximum of 6 ft when fully stretched.
- Flexible ducts typically have higher pressure losses than same-diameter metal ducts because of the increased roughness caused by the spiral helix inner liner. Liner compression will significantly increase the pressure loss. This needs to be accounted for by selecting a larger-size flexible duct that exhibits the same pressure loss as the corresponding metal duct. This can be done by using the flexible duct pressure drop correction factor (PDCF) equations from *ASHRAE Handbook—Fundamentals* (2017a), the ASHRAE/Air Distribution Institute (ADI) Duct Size Calculator (ASHRAE 2016b), or the ASHRAE Duct Fitting Database (DFDB; 2016a). DFDB fitting CD11-2 can be used to calculate the pressure drop of flexible duct for up to 20 in. diameter based on the installed length and compression. The DFDB fitting designations CD3-22 and CD3-23 can be used to evaluate 45° and 90° elbows for r/D = 1.0 and r/D = 1.5, respectively. In any of these cases, the DFDB cannot account for every situation of installed flexible duct. If the flexible duct is not properly installed, the pressure loss could be many magnitudes larger than the DFDB predicts. The installer should be trained and should follow the manufacturer's recommended installation procedures.

Textile Air Dispersion Systems

These systems are diffuser systems designed both to convey air within a room, space, or area and to diffuse air into that space while operating under positive pressure. These systems are commonly found in applications from open-ceiling architecture to underfloor air plenums. They should be listed and labeled according to UL 2518 (2016), which states that materials for air dispersion systems shall pass an erosion, high temperature, low temperature, pressure, mold growth and humidity, and surface burning characteristics test. The following materials are commonly available for this type of system:

- Woven polyester
- Woven and coated polyester
- Woven nonshedding polyester
- Antistatic polyester
- Antimicrobial-treated woven polyester
- Woven and coated polyethylene

ALTERNATIVE MATERIALS

Other materials may be used to convey air; however, they need to comply with the latest code requirements as follows:

- Have credible published and peer-reviewed absolute roughness factors in accordance with ANSI/ASHRAE Standard 120 (ASHRAE 2017b).
- Meet ASTM E84 (2020) and/or UL 181 (2013) flame and smoke requirements of 25/50.
- Meet the material strength characteristics of UL 181 (2013).
- Comply with ANSI/ASHRAE Standard 62.1 (ASHRAE 2019b).
- Be able to last for the lifetime of the intended use.

The material thickness, joining and support methods, installation guidelines and best practices, and sealing methods against air leakage are referenced by a credible and recognized organization.

REFERENCES

ACGIH. 2019. *Industrial ventilation: A manual of recommended practice for design*, 30th ed. Cincinnati, OH: American Conference of Governmental Industrial Hygienists.

ADC. 2010. *Flexible duct performance & installation standards*, 5th ed. Schaumburg, IL: Air Duct Council.

ASHRAE. 2016a. ASHRAE Duct Fitting Database, ver. 6.00.05 (online). Peachtree Corners, GA: ASHRAE. www.ashrae.org/technical-resources/bookstore/duct-fitting-database.

ASHRAE. 2016b. Duct Size Calculator. ASHRAE and Air Distribution Institute. Peachtree Corners, GA: ASHRAE.

ASHRAE. 2016b. *ASHRAE handbook—HVAC systems and equipment.* Peachtree Corners, GA: ASHRAE.

ASHRAE. 2017a. Chapter 21, Duct design. In *ASHRAE handbook—Fundamentals*. Peachtree Corners, GA: ASHRAE.

ASHRAE. 2017b. ANSI/ASHRAE Standard 120-2017, *Method of testing to determine flow resistance of HVAC ducts and fittings*. Peachtree Corners, GA: ASHRAE.

ASHRAE. 2018. ANSI/ASHRAE/IES Standard 90.2, *Energy-efficient design of low-rise residential buildings*. Peachtree Corners, GA: ASHRAE.

ASHRAE. 2019a. ANSI/ASHRAE/IES Standard 90.1, *Energy standard for buildings except low-rise residential buildings*. Peachtree Corners, GA: ASHRAE.

ASHRAE. 2019b. ANSI/ASHRAE Standard 62.1, *Ventilation for acceptable indoor air quality.* Peachtree Corners, GA: ASHRAE.

ASHRAE. 2019c. *ASHRAE handbook—HVAC applications*. Peachtree Corners, GA: ASHRAE.

ASTM. 2014. ASTM D7396-14, *Standard guide for preparation of new, continuous zinc-coated (galvanized) steel surfaces for painting*. West Conshohocken, PA: ASTM International.

ASTM. 2016a. ASTM D3273-16, *Standard test method for resistance to growth of mold on the surface of interior coatings in an environmental chamber*. West Conshohocken, PA: ASTM International.

ASTM. 2016b. ASTM D6386-16a, *Standard practice for preparation of zinc (hot-dip galvanized) coated iron and steel product and hardware surfaces for painting*. West Conshohocken, PA: ASTM International.

ASTM. 2019. ASTM C1338-19, *Standard test method for determining fungi resistance of insulation materials and facings*. West Conshohocken, PA: ASTM International.

ASTM. 20120. ASTM E84-19b, *Standard test method for surface burning characteristics of building materials*. West Conshohocken, PA: ASTM International.

De Renzo, D.J., and I. Mellan. 1985. *Corrosion resistant materials handbook*, 4th ed. Park Ridge, NJ: Noyes Data Corporation.

IAPMO. 2018. *Uniform mechanical code®*. Ontario, CA: International Association of Plumbing and Mechanical Officials.

ICC. 2018a. *International building code®*. Washington, DC: International Code Council.

ICC. 2018a. *International green construction code®* powered by ANSI/ASHRAE/ICC/USGBC/IES Standard 189.1. Washington, DC: International Code Council.

ICC. 2018b. *International mechanical code®*. Washington, DC: International Code Council.

NAIMA. 2016. Fibrous Glass Duct Construction Standards, 5th ed. Alexandria, VA: North American Insulation Manufacturers Association.

NFPA. 2015. NFPA 91, *Standard for exhaust systems for air conveying of vapors, gases, mists, and particulate solids*. Quincy, MA: National Fire Protection Association.

NFPA. 2018a. NFPA 90A, *Standard for the installation of air-conditioning and ventilating systems*. Quincy, MA: National Fire Protection Association.

NFPA. 2018b. NFPA 90B, *Standard for the installation of warm air heating and air-conditioning systems*. Quincy, MA: National Fire Protection Association.

SMACNA. 1995. *Thermoplastic duct (PVC) construction manual*, 2nd edition. Chantilly, VA: Sheet Metal and Air Conditioning Contractors' National Association.

SMACNA. 2000. *Duct cleanliness for new construction guidelines*. Chantilly, VA: Sheet Metal and Air Conditioning Contractors' National Association. https://www.smacna.org/technical/papers-and-guidelines/technical-paper/2000/05/27/duct-cleanliness-for-new-construction-guidelines.

SMACNA. 2001. *Kitchen ventilation systems and food service equipment fabrication and installation guidelines*. Chantilly, VA: Sheet Metal and Air Conditioning Contractors' National Association.

SMACNA. 2003. *Fibrous glass duct construction standards*. Chantilly, VA: Sheet Metal and Air Conditioning Contractors' National Association.

SMACNA. 2005. *HVAC duct construction standards—Metal and flexible*, 3rd ed. (ANSI/SMACNA 006-2006.) Chantilly, VA: Sheet Metal and Air Conditioning Contractors' National Association.

SMACNA. 2008a. *IAQ guidelines for occupied buildings under construction*, 2nd ed. (ANSI/SMACNA 008-2008.) Chantilly, VA: Sheet Metal and Air Conditioning Contractors' National Association.

SMACNA. 2008b. *Seismic restraint manual—Guidelines for mechanical systems*, 3rd ed. (ANSI/SMACNA 001-2008.) Chantilly, VA: Sheet Metal and Air Conditioning Contractors' National Association.

SMACNA. 2009. *Seismic restraint manual—Guidelines for mechanical systems—OSHPD edition*. (Developed to meet the specific requirements of the California Office of Statewide Health Planning and Development for hospital and health-care facilities construction.) Chantilly, VA: Sheet Metal and Air Conditioning Contractors' National Association.

SMACNA. 2011. *Rectangular industrial duct construction standards*, 2nd ed. (ANSI/SMACNA 002-2011.) Chantilly, VA: Sheet Metal and Air Conditioning Contractors' National Association.

SMACNA. 2013. *Round industrial duct construction standards*, 3rd ed. (ANSI/SMACNA 005-2013.) Chantilly, VA: Sheet Metal and Air Conditioning Contractors' National Association.

SMACNA. 2015. *Phenolic duct construction standards*. (ANSI/SMACNA 022-2015.) Chantilly, VA: Sheet Metal and Air Conditioning Contractors' National Association.

SMACNA. 2016. *Thermoset FRP duct construction manual*, 2nd ed. (ANSI/SMACNA 011-2017.) Chantilly, VA: Sheet Metal and Air Conditioning Contractors' National Association.

UL. 2013. UL 181, *Standard for factory-made air ducts and air connectors*, 11th ed. Northbrook, IL: UL LLC.

UL. 2016. UL 2518, *Standard for air dispersion systems*. Northbrook, IL: UL LLC.

Zinc. 2017a. GalvInfoNote 2.1, The continuous hot-dip coating process for steel sheet products. Durham, NC: International Zinc Association. https://www.galvinfo.com/wp-content/uploads/sites/8/2017/05/GalvInfoNote_2_1.pdf.

Zinc. 2017b. GalvInfoNote 2.11, Preparing galvanize for field painting. Durham, NC: International Zinc Association. https://www.galvinfo.com/wp-content/uploads/sites/8/2017/05/GalvInfoNote_2_11.pdf.

Zinc. 2017c. GalvInfoNote 2.12, Pretreatments for metallic-coated sheet. Durham, NC: International Zinc Association. https://www.galvinfo.com/wp-content/uploads/sites/8/2017/05/GalvInfoNote_2_12.pdf.

Zinc. 2017d. GalvInfoNote 3.2, Protecting galvanized steel sheet products from storage stain. Durham, NC: International Zinc Association. https://www.galvinfo.com/wp-content/uploads/sites/8/2017/05/GalvInfoNote_3_2.pdf.

Zinc. 2018. GalvInfoNote 1.3, Galvanneal – Differences from galvanized. Durham, NC: International Zinc Association. https://www.galvinfo.com/wp-content/uploads/sites/8/2017/05/GalvInfoNote_1_3.pdf.

Zinc. 2019. GalvInfoNote 1.1, understanding coating weight designations for zinc-based coating on steel sheet. Durham, NC: International Zinc Association. https://www.galvinfo.com/wp-content/uploads/sites/8/2017/05/GalvInfoNote_1_1.pdf.

BIBLIOGRAPHY

ASTM. 2013. ASTM B152/B152M-13, *Standard specification for copper sheet, strip, plate, and rolled bar*. West Conshohocken, PA: ASTM International.

ASTM. 2014. ASTM A36/A36M-14, *Standard specification for carbon structural steel*. West Conshohocken, PA: ASTM International.

ASTM. 2014. ASTM B209-14, *Standard specification for aluminum and aluminum-alloy and sheet and plate*. West Conshohocken, PA: ASTM International.

ASTM. 2014. ASTM C1139-14, *Standard specification for fibrous glass thermal insulation and sound absorbing blanket and board for military applications*. West Conshohocken, PA: ASTM International.

ASTM. 2014. ASTM D3982-08(2014), *Standard specification for contact molded "fiberglass" (glass fiber reinforced thermosetting resin) ducts*. West Conshohocken, PA: ASTM International.

ASTM. 2015. ASTM A463/A463M-15, *Standard specification for steel sheet, aluminum-coated, by the hot-dip process*. West Conshohocken, PA: ASTM International.

ASTM. 2015. ASTM C1696-15, *Standard test method for field measurement of raised retroreflective pavement markers using a portable retroreflectometer*. West Conshohocken, PA: ASTM International.

ASTM. 2016. ASTM C1290-16, *Standard specification for flexible fibrous glass blanket insulation used to externally insulate HVAC ducts*. West Conshohocken, PA: ASTM International.

ASTM. 2017. ASTM A568/A568M-17a, *Standard specification for steel, sheet, carbon, structural, and high-strength, low-alloy, hot-rolled and cold-rolled, general requirements for*. West Conshohocken, PA: ASTM International.

ASTM. 2017. ASTM C518-17, *Standard test method for steady-state thermal transmission properties by means of the heat flow meter apparatus*. West Conshohocken, PA: ASTM International.

ASTM. 2017. ASTM C635/C635M-17, *Standard specification for manufacture, performance, and testing of metal suspension systems for acoustical tile and lay-in panel ceilings*. West Conshohocken, PA: ASTM International.

ASTM. 2018. ASTM A1008/A1008M-18, *Standard specification for steel, sheet, cold-rolled, carbon, structural, high-strength low-alloy, high-strength low alloy with improved formability, solution hardened, and bake hardenable*. West Conshohocken, PA: ASTM International.

ASTM. 2018. ASTM A1011/A1011M-18a, *Standard specification for steel, sheet and strip, hot-rolled, carbon, structural, high-strength low-alloy, high strength low-alloy with improved formability, and ultra-high strength*. West Conshohocken, PA: ASTM International.

ASTM. 2018. ASTM A480/A480M-18, *Standard specification for general requirements for flat-rolled stainless and heat-resisting steel plate, sheet, and strip*. West Conshohocken, PA: ASTM International.

ASTM. 2018. ASTM A653/A653M-18, *Standard specification for steel sheet, zinc-coated (galvanized) or zinc-iron alloy-coated (galvannealed) by the hot-dip process*. West Conshohocken, PA: ASTM International.

ASTM. 2018. ASTM A924/A924M-18, *Standard specification for general requirements for steel sheet, metallic-coated by the hot-dip process*. West Conshohocken, PA: ASTM International.

ASTM. 2019. ASTM C1071-19, *Standard specification for fibrous glass duct lining insulation (thermal and sound absorbing material)*. West Conshohocken, PA: ASTM International.

ASTM. 2020. ASTM C1534-20a, *Standard specification for flexible polymeric foam sheet insulation used as a thermal and sound absorbing liner for duct systems*. West Conshohocken, PA: ASTM International.

UL. 2018. UL 723, *Standard for surface burning characteristics of building materials*, 11th ed. Northbrook, IL: UL LLC.

Zinc. 2019. GalvInfoNote 2.10, Imparting resistance to storage stain. Durham, NC: International Zinc Association. https://www.galvinfo.com/wp-content/uploads/sites/8/2019/01/GalvInfoNote-2-10-1.pdf.

8

Duct System Acoustics

OVERVIEW

What is noise? Noise is unwanted sound. Sound may become noise when it

- Is too loud (e.g., the sound is uncomfortable or makes speech difficult to understand)
- Contains unwanted pure tones (e.g., a whine, whistle, or hum)
- Contains unwanted information or is distracting (e.g., an adjacent telephone conversation or undesirable music)
- Is unexpected (e.g., the sound of breaking glass)
- Is uncontrolled (e.g., a neighbor's lawn mower)
- Happens at the wrong time (e.g., a door slamming in the middle of the night)
- Is unpleasant (e.g., a dripping faucet)
- Connotes unpleasant experiences (e.g., a mosquito buzz or a siren wail)
- Is any combination of the above descriptors

Understanding what noise is and how to control it is fundamental to duct design. This chapter covers fan noise, natural sound attenuation in duct systems, room acoustics, room design guidelines, and how to perform acoustical analyses of duct systems.

The Acoustical Fundamentals section of this chapter covers fundamentals such as acoustical rating systems, how sound is measured, radiated duct noise (breakout and break-in) and how to calculate breakout noise. This chapter does not cover mechanical equipment room sound isolation or vibration isolation and control; these topics are covered in Chapter 49, "Noise and Vibration Control," of *ASHRAE Handbook—HVAC Applications* (2019).

System noise calculations are not complicated. The procedure follows a logical and orderly sequence, and no special acoustical training is required. Repeated exposure to the various formulas, steps, and analyses will result in better understanding and interpretation of the design parameters that most significantly affect noise generation and attenuation. This understanding will allow designers to improve their duct designs in the early stages of projects to minimize unwanted noise.

Specifying quiet equipment and adding noise control materials to HVAC system designs are normal parts of the design process because they help control noise. To avoid noise problems, the air-handling unit (AHU) and ductwork should be designed with the best aerodynamics possible, which also reduces pressure drop and, therefore, fan energy requirements.

It is intended that Schaffer's *A Practical Guide to Noise and Vibration Control for HVAC Systems* (2005) be used in conjunction with this design guide. Schaffer's guide complements this guide by presenting practical design guidelines to help minimize the possibility of excessive HVAC system noise in and around buildings and by suggesting investigation methods to solve existing noise problems. Chapter 49, "Noise and Vibration Control," of *ASHRAE Handbook—HVAC Applications* (2019) is also an excellent reference for acoustical calculations.

ACOUSTICAL FUNDAMENTALS

Sound is a propagating disturbance in a fluid (air, gas, or liquid) or in a solid. Sound in air is called *airborne sound* or just *sound*. It is generated by a vibrating surface or turbulent airstream. Sound in solids is generally called *structure-borne sound*. In HVAC system design, both airborne and structure-borne sound are important considerations.

Sound Pressure

Sound pressure is a fluctuation of the ambient air pressure generated by a sound source. These pressure variations are what our ears perceive and sound meters measure. Sound pressure is expressed in the pressure unit of Newtons per square meter (N/m^2) or Pascals (Pa). The actual measured sound pressure will depend on several factors, including the magnitude of the source of the sound, the pressure measurement location with respect to the source, and the conditions along the propagation path from the source to the measurement location. For example, if a person were half a mile away from a jet engine, it likely would not to be very loud, especially if the person is in a car with the windows rolled up. However, if a person were right beside the jet engine, the noise would be overwhelming.

The range of sound pressure magnitudes that humans can hear without severe hearing damage is very large. Sound pressure can range from 0 dB (0.00002 Pa, often written as 20 μPa) for excellent youthful hearing to 120 dB (20 Pa for passenger jet takeoff at 50 ft). That range represents a pressure ratio of about 1 to 1 million.

Sound Power

Sound power is sound energy. It is possible to convert between sound power from a piece of equipment to sound pressure at the location of a listener if there are enough data. A close analogy is that sound power is like lighting lumens from a lamp and sound pressure is like the foot-candles incident at a given location. There are many conditions that affect the sound pressure level that we hear from a source that creates sound power (magnitude of the source, path the sound takes, age of our ears). Therefore, rating sources of sound in terms of sound pressure is complex, because the sound pressure rating is only good for those exact conditions (e.g., 80 dB at 33 ft in a hemispherical free field). If a person were to stand farther away from that source or put a barrier between themselves and that source, the sound waves that hit their eardrums would be different and would be perceived as a different sound. To determine how sound pressure is affected under varying conditions, sound sources are most effectively rated in terms of acoustical energy, or sound power, which is expressed in terms of watts of sound power.

The sound power magnitudes also have a very large range (see Table 8-1). For example, the sound power of a very faint noise of 0 dB of sound power is 0.000,000,000,001 (1×10^{-12}) W. Quieter sources of sound would have negative sound power levels, but these hold no interest in the HVAC industry. The sound power generated by a space shuttle launch is on the order of 100,000,000 (1×10^{8}) W.

THE DECIBEL

The sound pressure level (L_p or SPL) describes the sound that is heard (or the loudness level of the sound) and is analogous to the foot-candle level at a particular location. This value varies with

Table 8-1 Typical Sound Power Outputs and Sound Power Levels
(Reproduced from ASHRAE 2017, Table 2)

Source	Sound Power, W	Sound Power Level*, dB re 10^{-12} W
Space shuttle launch	10^8	200
Jet aircraft at takeoff	10^4	160
Large pipe organ	10	130
Small aircraft engine	1	120
Large HVAC fan	0.1	110
Heavy truck at highway speed	0.01	100
Voice, shouting	0.001	90
Garbage disposal unit	10^{-4}	80
Voice, conversation level	10^{-5}	70
Electronic equipment ventilation fan	10^{-6}	60
Office air diffuser	10^{-7}	50
Small electric clock	10^{-8}	40
Voice, soft whisper	10^{-9}	30
Rustling leaves	10^{-10}	20
Human breath	10^{-11}	10

*Calculated per Equation 8-2.

the distance from the sound source and the environment surrounding the sound source. Sound pressure is expressed in decibels (dB) with a reference level to 0.00002 Pa:

$$L_p = 20\log_{10}\left(\frac{p}{p_{ref}}\right) \tag{8-1}$$

The sound power level (L_w or PWL) describes the total amount of acoustical energy and is analogous to a light bulb rating of 1000 lumens (lm). This value is independent of location, distance, and environment. Sound power is expressed in decibels (dB) with a reference level of 10^{-12} W:

$$L_w = 10\log_{10}\left(\frac{w}{w_{ref}}\right) \tag{8-2}$$

A good mnemonic for these equations is that p stands for pressure and w stands for watts or power.

Because the human ear can measure over a range of about 120 dB (12 zeros), decibels are the primary unit of sound measurement used to quantify both sound pressure level and sound power level. The ranges of sound pressures and sound powers are so large that the magnitudes of sound physical properties are expressed in decibels (dB). Decibels show a ratio between items on a logarithmic scale. As shown in Equations 8-1 and 8-2, the level L is based on the common (base 10) logarithm of a ratio of the magnitude of a physical property of pressure or power to a reference

Table 8-2 Subjective Reactions to Ambient Sound Pressure Levels (Schaffer 1991)

Change from Ambient Level	Subjective Reaction
1 dB	Not perceptible
2–3 dB	Just perceptible
4–5 dB	Clearly perceptible
9–10 dB	Perceived as twice as loud

magnitude of the same property. In these equations, a factor of 10 is included to convert bels to decibels. The *bel* originates from Alexander Graham Bell, who invented the bel scale, but it sometimes resulted in very small numbers, so today they are multiplied by 10, thus *decibel.* Multiplying by 10 makes the math more complex but avoids fractional numbers in almost all cases. The decibel scale is the preferred method of presenting quantities in acoustics, not only because it collapses a large range to a more manageable range but also because its levels correlate better with human responses to the magnitude of sound than do absolute numbers for sound pressure. Human ears are sensitive to orders of magnitude changes (multiples of 10), not changes of just a few Pascals or watts.

Because sound pressure levels are logarithmic quantities, an increase or decrease of only a few decibels is not significant. For example, a sound pressure level change of 3 dB represents a *mathematical* doubling (or halving) of the sound pressure, but human ears will not detect this change as a doubling or a halving of the subjective loudness of the sound: if the sound pressure is 2 Pa then the sound pressure level is 100 dB, but if the sound pressure doubles to 4 Pa then the sound pressure level only increases to 103 dB. Human reactions to changes in sound level are summarized in Table 8-2.

Most people have some familiarity with sound pressure levels measured in decibels from listening to music or because they must comply with Occupational Safety and Health Administration (OSHA) regulations during the course of their jobs. Current OSHA regulations (OSHA n.d.) require actions by employers when continued exposure to sound pressure levels more than 85 dBA exist, because that level of sound pressure can result in hearing impairment. Any exposure to levels above 140 dBA can result in rapid permanent hearing damage.

COMBINING SOUND PRESSURE LEVELS

Because sound levels that are expressed in dB notation are logarithmic quantities, they cannot be added directly. For example, two noise sources of 100 dB each do not equal 200 dB. The combined level from these sources is 103 dB determined using Equation 8-3:

$$L_p = 10\log_{10}(10\wedge(\text{dB1}/10) + 10\wedge(\text{dB2}/10) + \text{etc.}) \tag{8-3}$$

In this case,

$$L_p = 10\log_{10}(10\wedge(100\text{ dB}/10) + 10\wedge(100\text{ dB2}/10)) = 103\text{ dB}$$

With the number of sound pressure levels that must be combined in a duct system, the calculations are best done using computers. However, the rules of thumb in this section are helpful for making fairly accurate manual calculations. Any number of sound levels can be combined by determining the difference between two of the levels and adding the adjustments shown in Table 8-3 to the higher of the two levels. Whenever the difference between two sound levels is 10 dB or more, the

Table 8-3 Combining Two Sound Levels

Difference in Levels, dB	Add to Higher Level
0 to 1	3
2 to 4	2
5 to 9	1
10 or more	0

Table 8-4 Combining Identical Sound Levels

Number of Identical L_p or L_w Sounds	Add to Level
2	3
10	10
100	20
1,000	30

louder level masks the quieter source and therefore there is virtually no contribution to the overall level by the lower source.

Another rule of thumb to remember when adding *identical* sound levels, such as an array of fans or diffusers, in the same location or at the same distance from the listener is to add the adjustments shown in Table 8-4.

Example 8-1 shows how to determine overall sound pressure level.

FREQUENCY

The frequency of sound is determined by the number of oscillations (or cycles) completed per second by a vibrating object. Frequency is measured in cycles per second or hertz (Hz). The audible frequency range for humans with unimpaired hearing extends from about 20 to 20,000 Hz. As a point of reference, middle C on a piano keyboard is 262 Hz, which is close to the center frequency of the third octave band, which is 250 Hz. Moving upward, each octave band has a center frequency twice the value of the lower octave center frequency. Similarly, moving downward, each octave band has a center frequency one-half that of the upper octave. This relationship is true regardless of which frequency is selected as a starting point.

The standard eight octave bands used in duct system acoustics are shown in Table 8-5.

Many noise problems, including those in HVAC systems, must be analyzed as a function of frequency. Due to the differences in their wavelengths, high-frequency sounds (short wavelengths) can behave quite differently from low-frequency sounds (long wavelengths). Common noise problems from fans often occur in the 125 to 250 Hz range. Noise levels in the 16 and 31.5 Hz bands are generally not problems but could generate low-frequency rumble under certain situations.

LOUDNESS

Loudness can be defined as the perceived volume of sound and is a function of both sound pressure level and frequency. Various metrics have been created to measure loudness, but all recognize that low-frequency noise is more tolerable than similar levels at higher frequencies.

Example 8-1.
A listener is subjected to the following sound pressure levels: 51 dB, 53 dB, 49 dB, 45 dB, 36 dB, 31 dB, 25 dB, and 24 dB. What is the overall sound pressure level?

Solution.
The sound pressure levels are added in groups of two in accordance with Table 8-3, and the results of these groups are then coupled in like manner until a single sound level is attained (see "Approximate Solution" below). The additions can be carried out in any order and the results should be identical (or should vary by no more than 1 dB). Using this rule of thumb, the overall sound pressure level is 57 dB. This is considered the approximate solution because using Equation 8-3 results in an overall sound pressure level exact solution of 56.5 dB (see "Exact Solution" below). However, the difference between the two results is only a trivial difference in acoustics.

Approximate Solution Using Table 8-3

(51 + 53) dB = 53 + 2 = 55 dB

(49 + 45) dB = 49 + 2 = 51 dB

(36 + 31) dB = 36 + 1 = 37 dB

(25 + 24) dB = 25 + 3 = 28 dB

Add these results together:

(55 + 51) dB = 55 + 2 = 57 dB

(37 + 28) dB = 37 + 1 = 38 dB

Finally, (57 + 38) dB = 57 + 0 = **57 dB,** the overall level.

Exact Solution Using Equation 8-3

	dB	10^(dB/10)
dB1	51	125893
dB2	53	199526
dB3	49	79433
dB4	45	31623
dB5	36	3981
dB6	31	1259
dB7	25	316
dB8	24	251
	sum 10^(dB/10) =	442282
	10*Log$_{10}$(sum 10^(dB/10)) =	**56.5 dB**

Based on numerous surveys conducted with a wide range of human subjects, equal loudness contours have been created that provide an indication of the actual pure tone sound pressure levels at various frequencies that are judged to be equal in loudness to a reference tone at 1000 Hz (see Figure 8-1). For example, in the first octave band (63 Hz), a 61 dB tone is considered to have the same loudness as a 40 dB tone at 1000 Hz. In the seventh octave band (4000 Hz), a 33 dB level is judged to be equal to the same 40 dB tone at 1000 Hz. Thus, the frequency of a sound has a substantial bearing on how loud it is perceived to be. The shapes of the curves are much different for quiet sounds than for loud sounds. One application of this phenomenon is the "loudness" or "bass boost" button on many amplifiers. It is intended to make music that is played at low volumes sound as it was intended at concert volume levels. Amplification of the low frequencies (bass) is always increased compared to the mid-frequencies, and in higher-quality equipment the highest frequencies also have increased amplification. Human ears are closer to being linear transducers at high volumes.

ACOUSTICAL RATING SYSTEMS

Human responses to noise depend on many factors (such as frequency content, level, and repetition rate), so it is difficult to define acceptable noise environments. Many rating systems have been developed to describe acceptable sound exposures in and around buildings. Each system has a

Table 8-5 Octave Bands

Band No.	—	—	1	2	3	4	5	6	7	8	—
Center frequency, Hz	16	31.5	63	125	250	500	1000	2000	4000	8000	16,000
Wavelength of center frequency, ft	71.4	36.3	18.3	9.1	4.6	2.3	1.1	0.58	0.28	0.14	0.07
Range, Hz	11.2 to 22.4	22.4 to 45	45 to 90	90 to 180	180 to 355	355 to 710	710 to 1400	1400 to 2800	2800 to 5600	5600 to 11,200	11,200 to 22,400

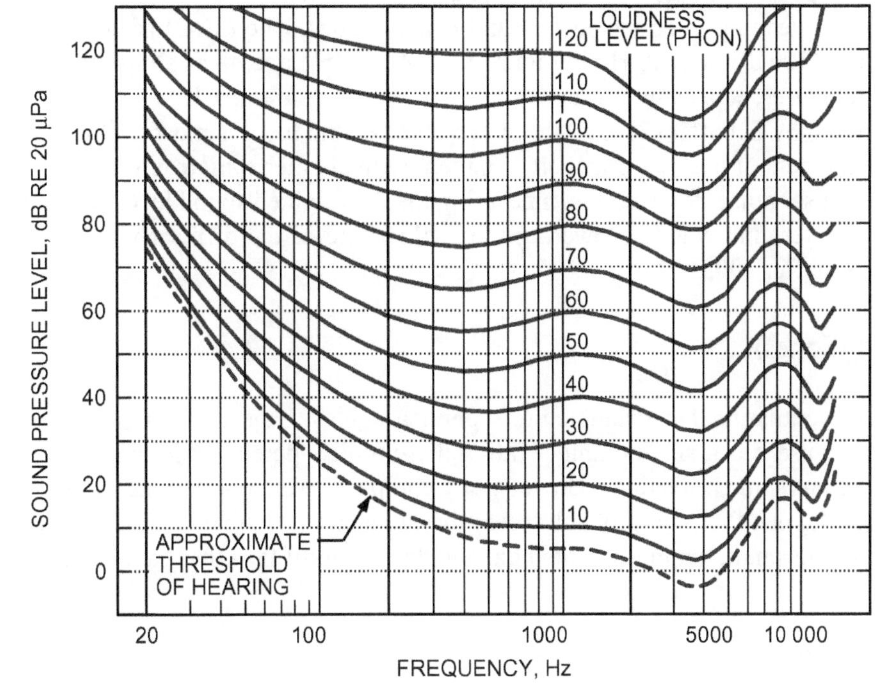

Figure 8-1 Equal Loudness Contours
(Reprinted from ASHRAE 2017, Figure 4)

set of assumptions and limiting conditions (Schaffer 2005). The rating systems most frequently used to describe HVAC system noise are the following:

- A-weighted sound level—dBA, sometimes written "dB(A)"
- Loudness level (sones)
- Noise Criteria (NC)
- Room Criteria (RC) Mark II

A-Weighted Sound Level—dBA or dB(A)

For broadband sounds at low loudness, the A-weighted sound level indicates approximate relative loudness as perceived by persons and is measured using the A-weighting network of a sound level meter. It is a single number that is used to rate the perceived volume of noise and that is commonly used in the areas of hearing conversation and community noise exposure. It has gained popularity because it is easily measured with an inexpensive sound level meter and is a good predictor of how loud (not how annoying) persons will perceive a sound to be (Schaffer 2005). However, this rating system understates the annoyance level of low-frequency sounds, such as those associated with duct rumble, and therefore may not properly identify low-frequency noise problems. As for all single-number sound metrics, a single number cannot indicate whether noise problems are in the lower, middle, or upper frequencies, the identification of which is critical to determining how to prevent or solve acoustical problems. The dBA curve approximates the equal loudness contour (see Figure 8-1) with 40 dB of sound pressure at 1000 Hz (also known as the *40 phon curve*). To determine the dBA, the decibels in each octave are adjusted according to the negative of the curve in Figure 8-2 and the octave readings are then added logarithmically. For example, 39.4 dB is subtracted from the 31.5 Hz octave, while the 1000 Hz octave is not adjusted before adding the octave bands.

Rating systems using B-weighted and C-weighted sound levels, dB(B) and dB(C), have also been developed for louder sounds, though dB(B) is almost never used. Some sound meters have both "flat" (meaning uncorrected dB) and dB(C) scales. The dB(C) scale approximates the 100 phon curve (100 dB at 1000 Hz in Figure 8-1). As people age, their ability to hear high-frequency sounds decreases, especially for men.

As noted by Schaffer (2005), "the A-weighted sound level is most useful when comparing the relative loudness of one acoustical environment with another *similar* environment" (p. 177). In the case of diffuser noise, for example, measuring the A-weighted sound level at different airflow rates may help determine whether lowering the airflow rate of the diffuser will be effective for noise control (Schaffer 2005).

Loudness Level (Sones)

Many ventilating fan manufacturers publish sone ratings of their fans at various operating points. Because it is a single-number rating that predicts perceived loudness (not annoyance), the sone value has some limitations. However, by comparing the sone ratings of similar fans, designers can select the one that will probably seem the quietest to occupants. Sone rating is mainly useful for comparing small, ceiling-mounted exhaust fans. For all fans that have ductwork between them

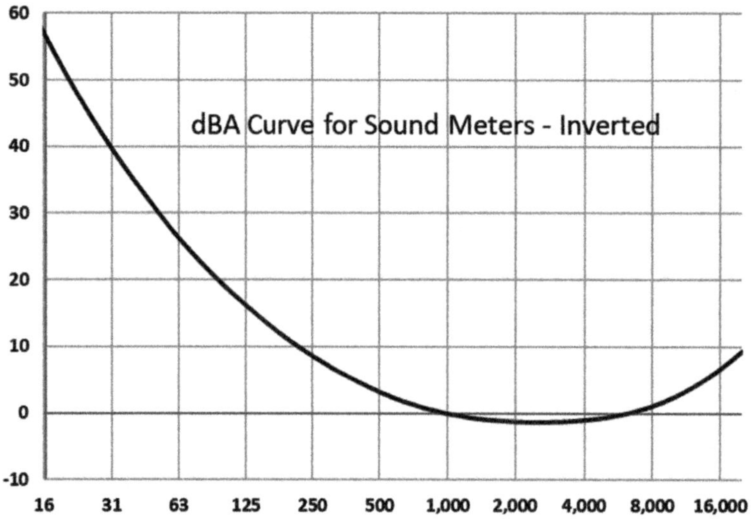

Figure 8-2 dBA Curve, Inverted

and the occupied spaces, using more complete octave band data instead of sones is recommended. The value of 1 sone is the perceived noise level of a 1000 Hz sound with a sound pressure level of 40 dB. Sone ratings, unlike measures that use dB, have a doubling of the sone rating, which means that the average person will perceive the sound as twice as loud as it actually is. Doubling of the sone rating is equivalent to a 10 dB increase in sound level.

Noise Criteria (NC)

The Noise Criteria (NC) rating system uses rating curves somewhat like those in the dBA system but is intended to indicate annoyance more than loudness. Variations of NC ratings have also been developed, such as Preferred Noise Criteria (PNC) and Balanced Noise Criteria (NCB). NC, PNC, and especially NCB are precursors to the Room Criteria (RC) rating system. Unfortunately, many designers are not familiar with the PNC or RC noise criteria, so they are not as widely used as NC. The NC rating is found by determining the highest penetration of any octave band level into the NC curves. In the example illustrated in Figure 8-3, the octave band penetrating the highest into the NC curves is 45 dB in the 125 Hz octave band. In this case, the noise spectrum is given a rating of NC-45.

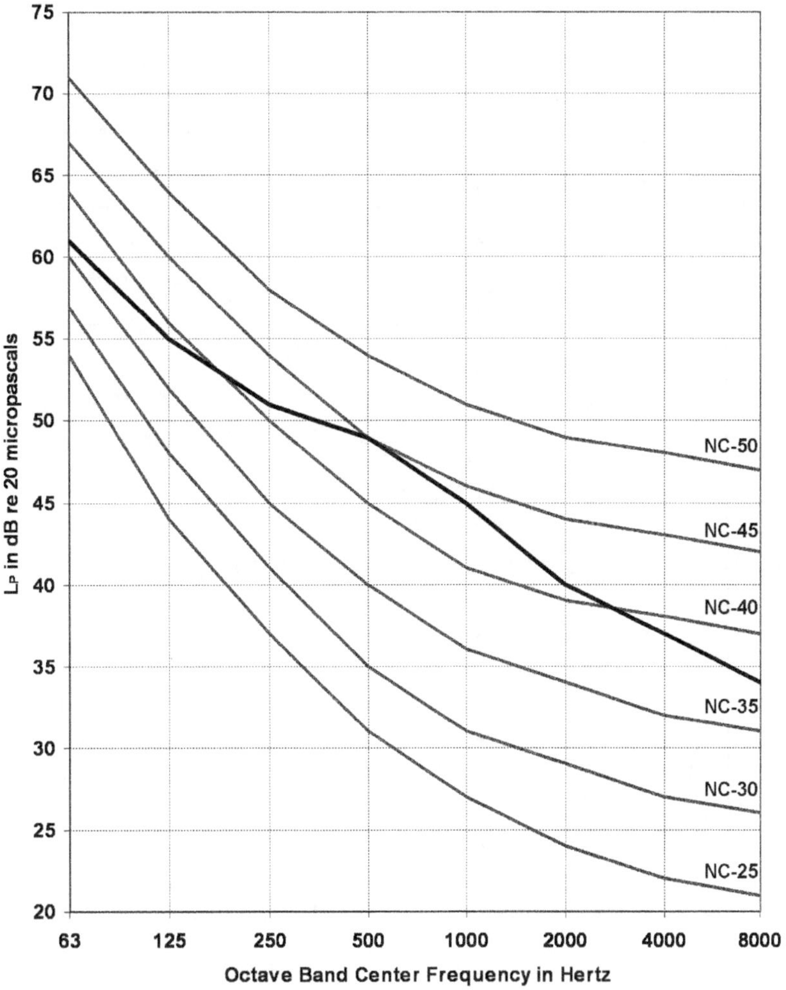

Figure 8-3 Application of the NC Rating System to a Measured Spectrum
(Reprinted from Schaffer 2005, Figure B-5)

This system is widely used in the HVAC industry; however, the NC rating system should be used with caution because of the following limitations:

- NC curves do not extend to the 16 and 32 Hz octave bands, which are the bands into which some of the most problematic HVAC noise issues can fall (Schaffer 2005). The NCB method does extend to these bands.
- By reporting a single number based on a tangency to a rating curve at one octave band without assigning a subjective quality to the sound spectrum, one may mistakenly rate two widely different sound spectra as equally acceptable. Figure 8-4 shows an example of this wherein a fan noise spectrum that is dominated by low-frequency noise could have the same NC rating as a grille noise spectrum that is dominated by high-frequency noise.

The NCB method is an improvement to the NC rating system that takes into account the 16 and 32 Hz octave bands. It led to the development of RC and RC Mark II, which are each single-number-plus-modifier noise rating systems to diagnose and rate the HVAC noise exposure in a room. The NCB method provides both an RC rating and a modifier to provide more information than a single NC number, especially for low frequency noise and noise character.

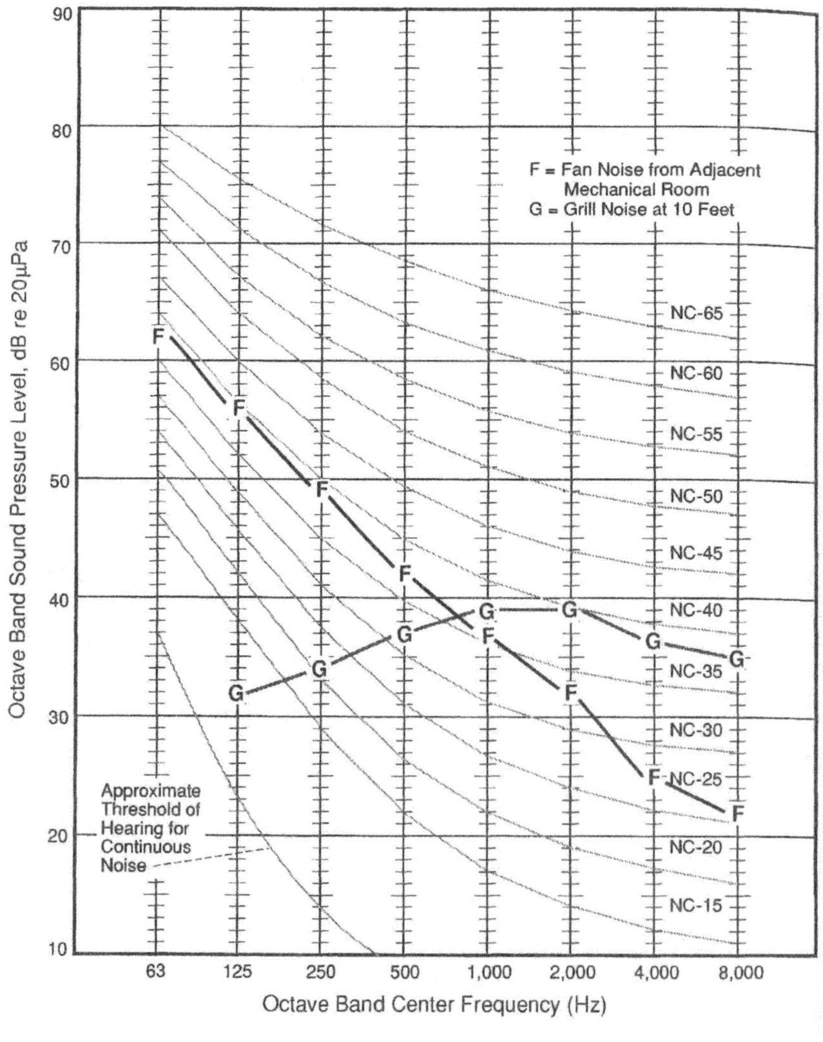

Figure 8-4 Two Different HVAC Noise Spectra with the Same NC Rating
(Reprinted from Schaffer 1991, Figure B–3)

Room Criteria (RC) Mark II

For duct design sound spectrum calculation, the Room Criteria (RC) method is recommended. The RC rating of a sound environment has two factors: a numerical value that rates the ease of voice communication and a letter suffix indicating the subjective quality of the noise (Schaffer 2005). Because RC includes two factors, it provides more information than single-number metrics such as dBA, sones, and NC. RC Mark II modified the RC method in 1997 and is now recommended by ASHRAE (2019). The RC rating system was originally designed as a diagnostic procedure to rate a measured sound spectrum. Since its introduction it has also been used as a design tool for predictive purposes.

For diagnostic tests, the following steps are used to determine the complete RC rating of a sound spectrum. The RC rating for the plotted data from 16 to 4000 Hz in Figure 8-2 is RC-37R, where "R" indicates "rumble."

1. Measure the sound level in the nine octave bands from 16 to 4000 Hz.
2. Plot the nine octave band sound levels on a blank RC work chart (Figure 8-5), with lines connecting the plotted points (Figure 8-6) and analysis (Figure 8-7).

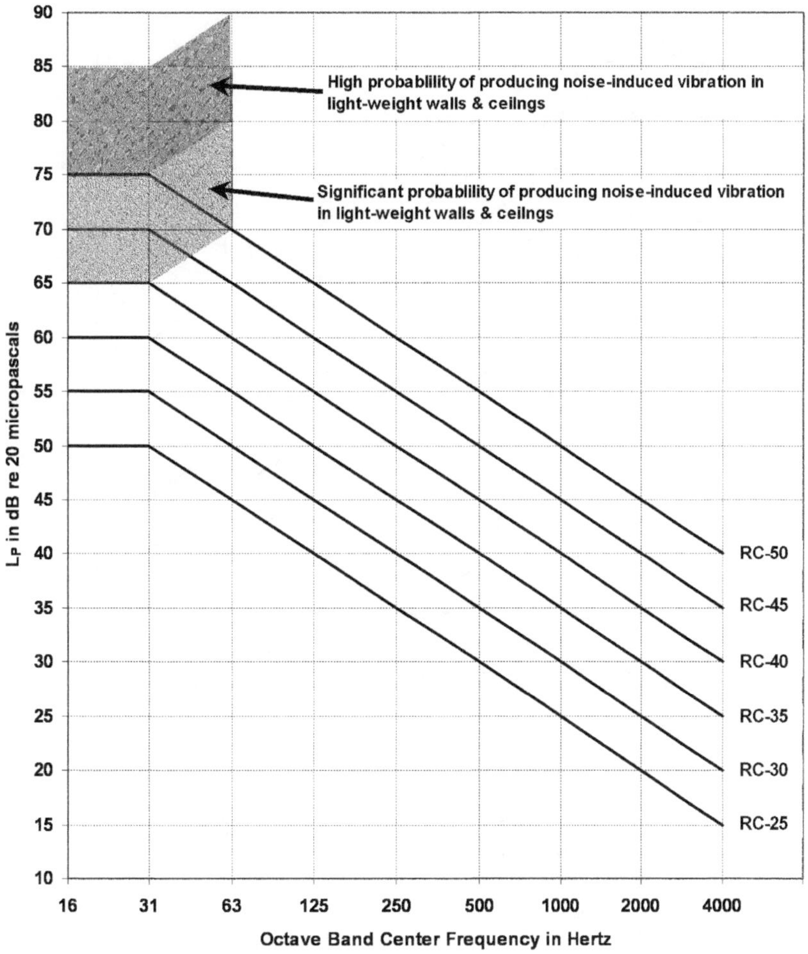

Figure 8-5 Blank RC Mark II Work Chart
(Reprinted from Schaffer 2005, Figure B-3)

3. Determine the arithmetic (not logarithmic) average of the 500, 1000, and 2000 Hz octave band sound levels. The result is the numerical RC value. Figure 8-6 shows an example of points plotted on an RC work chart. The resultant RC value is 37 (Figure 8-7).
4. Draw a line on the RC chart that passes through the 1000 Hz column at the numerical RC value (37) and slopes at 5 dB per octave band, flattening below the 31 Hz octave (i.e., the values at 16 and 31 Hz are identical).
5. Determine the low-frequency (LF), medium-frequency (MF), and high-frequency (HF) mean sound pressure levels. The appropriate formula is LF = 10*log[((10^(0.1*ΔL_{16}) + 10^(0.1*$\Delta L_{31.5}$) + 10^(0.1*ΔL_{63}))/3] from Chapter 49, "Noise and Vibration Control," of *ASHRAE Handbook—HVAC Applications* (2019). Substitute MF and HF for the other two frequency ranges.
6. The difference between the highest and lowest of the items calculated in step 5 is the quality assessment index (QAI) value. If the QAI value is >5, the RC rating must include either

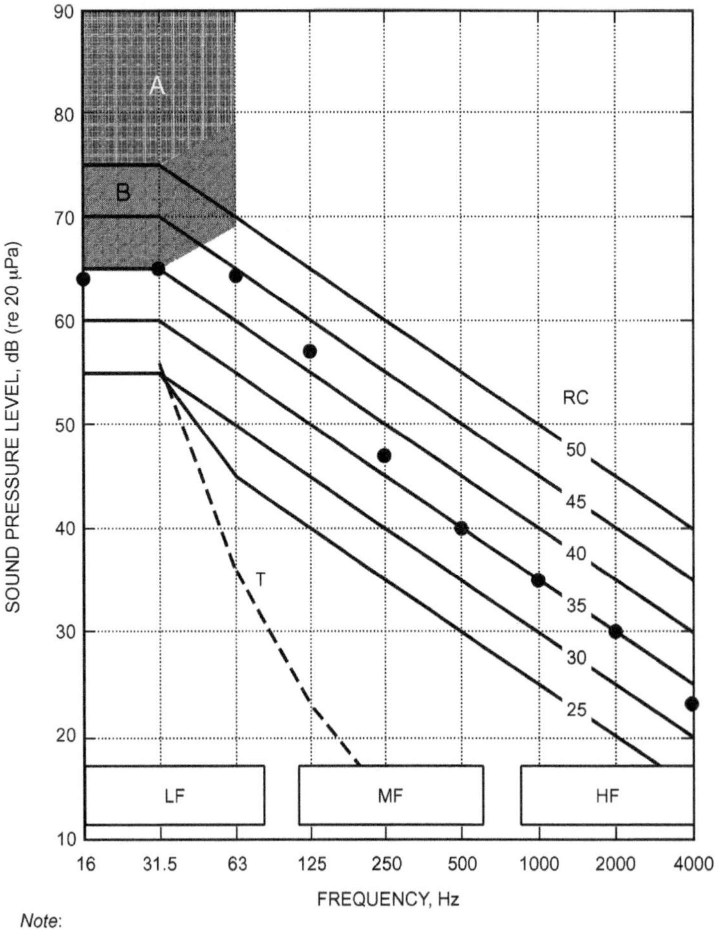

Note:
- Noise levels for lightweight wall and ceiling constructions:
  - In shaded region B are likely to generate vibration that may be perceptible. There is a slight possibility of rattles in light fixtures, doors, windows, etc.
  - In shaded region A have a high probability of generating easily perceptible noise-induced vibration. Audible rattling in light fixtures, doors, windows, etc. may be anticipated.
- Regions LF, MF, and HF are explained in the text.
- Solid dots are sound pressure levels for the example discussed in the text.

Figure 8-6 RC Mark II Work Chart with Plotted Points
(Reprinted from ASHRAE 2019, Figure 6)

LF, MF, or HF, whichever is highest. If the QAI value is <5, the assigned designation is N, which stands for *neutral.*

7. Sound levels in Region B (shown in Figure 8-7) may generate perceptible vibration in light wall and ceiling constructions. Rattles in light fixtures, doors, windows, etc., are a slight possibility.
8. Sound levels in Region A (shown in Figure 8-7) have a high probability of generating perceptible sound-induced vibration to light walls and ceiling constructions. Audible rattling in light fixtures, doors, windows, etc., should be anticipated.

Note that the RC rating system includes criteria for the 16 Hz octave band, where low-frequency problems can occur (see Figures 8-6 and 8-7).

For design calculations, the calculated octave band spectrum is from 16 to 4000 Hz. The RC rating for the plotted data in Figure 8-6 from 31.5 to 4000 Hz is RC-37R, with the slight possibility that light fixtures, doors, windows, etc., may rattle.

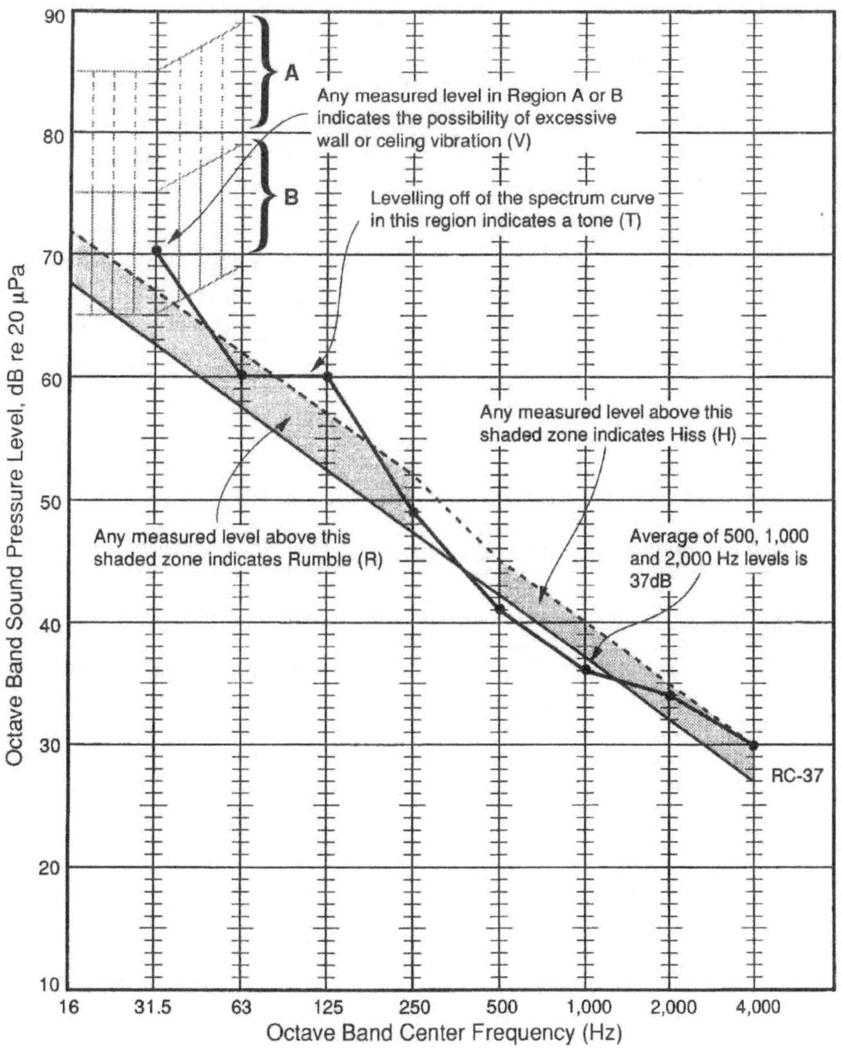

Figure 8-7 Application of the RC Rating System
(Reprinted from Schaffer 1991, Figure B–1)

MEASURING HVAC SYSTEM NOISE

Most HVAC sound level measurements can be made using a handheld, battery-operated sound level meter, which has a microphone, internal electronic circuits, and a readout display to measure and display the sound pressure level at the desired location. The most useful meters have octave band filters with center frequencies from 16 to 8000 Hz, linear and A and C weighting networks, high (Type 1) or moderate (Type 2) precision, and a digital display to show sound level fluctuations (Schaffer 2005). Sound level meters that do not have octave band filters are far less useful because they do not provide much help when trying to determine the frequency content of an offending noise, a necessary step in determining noise-reduction measures (Schaffer 2005). For less critical measurements, there are several apps for modern cellular phones that can perform well, although phones are not certified as Type 1 or Type 2 instruments in terms of precision. Also, phone cases have been found to affect reading accuracy.

Troubleshooting noise problems requires measuring the sound pressure level values in the 31 to 4000 Hz octave bands, and sometimes in the 16 Hz band (Schaffer 2005). For some limited cases, a sound level meter's linear, A, and C scales are useful if the meter cannot measure octave band sound levels. Figure 8-8 compares low-frequency sound from the three scales.

To investigate a rumble problem, compare the sound level measurements using the linear and C or flat networks. A 5 dB or more measurement difference might indicate that sound at frequencies less than 50 Hz is the problem. A similar comparison of measurements using the C and A scales can help determine the relative level of sound below 500 Hz, where the A-weighting network filter has the greatest effect. If the difference between C- and A-weighted levels for a sound exceeds about 20 dB, then the sound is likely to be annoying because of excessive low-frequency noise.

Frequently, a sound level meter readout will show a fluctuating value, especially where low-frequency noise is significant, because sound levels are rarely constant. If it fluctuates less than 4 dB, then record the "eyeball" average; if it fluctuates 5 dB or more, record both the highest and the lowest displayed values (Schaffer 2005). For a blank field data sheet and an example of a completed one, refer to Figures C-3 and C-4 of *A Practical Guide to Noise and Vibration Control for HVAC Systems* (Schaffer 2005).

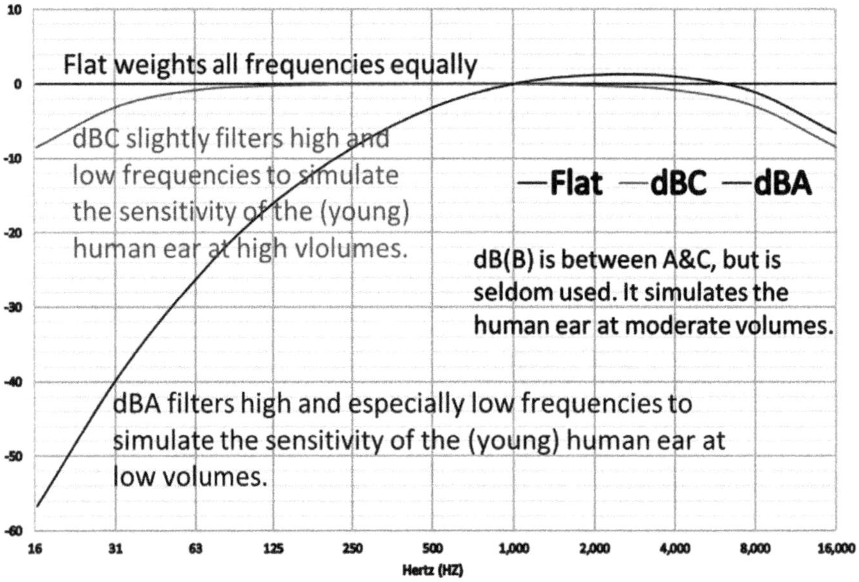

Figure 8-8 Frequency Weighting Effects of Sound Level Meter Weighting Networks

RADIATED DUCT NOISE

Breakout Noise

Noise that is within a duct, such as fan noise or generated noise, and that transmits through the duct wall into the surrounding area is called *breakout noise*. This phenomenon is often referred to as *low-frequency duct rumble*. There are two possible sources for duct breakout noise. One is associated with noise that is generated by fans. This noise is transmitted down the duct and then through the duct walls into surrounding spaces. The second source is associated with turbulent airflow that excites the duct walls, causing them to vibrate. This vibration can generate low-frequency duct rumble, which is then radiated into surrounding spaces. In many situations duct breakout noise may be associated with both sources. Breakout noise can be a problem if it is not adequately attenuated before the duct crosses over or under an occupied space. The main factors affecting breakout sound transmission are the sound power level in the duct at a location, transmission loss of the duct, total exposed surface area of the duct, and presence of any acoustical duct liner.

Transmission loss (TL) is the ratio of sound power incident on a partition (duct wall) to the sound power transmitted through the partition. This ratio varies with acoustic frequency as well as duct shape, size, wall thickness, pressure class, and reinforcement. Higher values of transmission loss result in less noise passing through the duct wall.

The sound power that is transmitted through a duct wall and then radiated from the exterior surface of the duct wall is given by Equation 8-4:

$$L_{w(out)} = L_{w(in)} + 10\log_{10}\left(\frac{S}{A}\right) - \mathrm{TL}_{out} \tag{8-4}$$

Calculating the sound power at the location of interest requires a bit of work. Several manufacturers' programs can calculate this beginning with the sound power of the fan discharge and working down the duct system to the critical location. Equation 8-4 assumes no decrease in the internal sound power level with distance along the length of the duct. Thus, it is valid only for relatively short lengths of unlined duct. For long ducts or ducts that have internal acoustic lining, one approach is to divide the duct into sections, each of which is short enough to be modeled as a section of duct with constant internal sound power level over the length of each section. For breakout noise, use Equation 8-5. This equation requires the calculation of effective duct length L^*, where L^* is the length that results in a 1 dB reduction in the internal sound power level at the frequency of interest. Equation 8-6 can be used for unlined and lined duct.

$$L_{w(out)} = L_{w(in)} + 10\log_{10}\left(\frac{S^*}{A}\right) - \mathrm{TL}_{out} \tag{8-5}$$

where

$$S^* = PL^* \tag{8-6}$$

The effective duct length is calculated as

$$L^* = \frac{\gamma^L - 1}{\ln\gamma}$$

where

$$\gamma = 10^{((-\alpha)/10)}$$

The quantity α denotes the duct attenuation rate as measured in units of dB/ft. Equations for S^* and A for rectangular ducts are as follows:

$$S^* = (2)(L^*)\left(\frac{W+H}{12}\right)$$

$$A = \frac{WH}{144}$$

Equations for S^* and A for round ducts are as follows:

$$S^* = L^*\pi D$$

$$A = \pi\left(\frac{D^2}{4}\right)$$

Values of TL_{out} for rectangular ducts are given in Table 8-6 and for round ducts in Table 8-7. Table 8-8 gives estimates of attenuation of different ceiling constructions by octave band. These data are used for estimating noise reduction between the ductwork and the receiver (the person or thing that perceives the noise).

In rooms with exposed ductwork, an estimate of the breakout sound pressure level at a specific point in space can be obtained using Equation 8-7:

$$L_p' = L_{w(out)} - 10\log_{10}(\pi rL) + 10 \qquad (8\text{-}7)$$

Example 8-2 shows how to determine breakout sound power level.

Break-In Noise

For details on how to calculate break-in noise from the surroundings to a duct, consult Chapter 49, "Noise and Vibration Control," of *ASHRAE Handbook—HVAC Applications* (2019).

Table 8-6 TL_{out} Versus Frequency for Rectangular Ducts
(Reproduced from ASHRAE 2019, Table 29)

Duct Size, in.	Gage	TL_{out}, dB							
		Octave Band Center Frequency, Hz							
		63	125	250	500	1000	2000	4000	8000
12 × 12	24	21	24	27	30	33	36	41	45
12 × 24	24	19	22	25	28	31	35	41	45
12 × 48	22	19	22	25	28	31	37	43	45
24 × 24	22	20	23	26	29	32	37	43	45
24 × 48	20	20	23	26	29	31	39	45	45
48 × 48	18	21	24	27	30	35	41	45	45
48 × 96	18	19	22	25	29	35	41	45	45

*Ducts internally lined with 1 in. thick 1.5 pcf fiberglass with 24 gage perforated sheet metal inner liner.

Table 8-7 TL_{out} versus Frequency for Round Ducts
(Reproduced from ASHRAE 2019, Table 30)

Diameter, in.		Gage	TL_{out}, dB						
			Octave Band Center Frequency, Hz						
			63	125	250	500	1000	2000	4000
Longitudinal Seam Duct									
8	15	26	>45	53	55	52	44	35	34
14	15	24	>50	60	54	36	34	31	25
22	15	22	>47	53	37	33	33	27	25
32	15	22	51	46	26	26	24	22	38
Spiral Duct									
12	12	26*	52	51	53	51	50	46	36
24	24	24	51	53	51	44	36	26	29
	24	24*	51	51	54	44	39	33	47
	10	16	>48	53	36	32	32	28	41
36	24	20	51	51	52	46	36	32	55

Table 8-8 Ceiling/Plenum/Room Sound Power Attenuation in Decibels for Generic Ceiling in T-Bar Suspension Systems
(Reproduced from ASHRAE 2019, Table 43)

Tile Type	Approx. Density, lb/ft^3	Tile Thickness, in.	Octave Band Center Frequency, Hz						
			63	125	250	500	1000	2000	4000
Mineral fiber	1.0	5/8	13	16	18	20	26	31	36
	0.5	5/8	13	15	17	19	25	30	33
Glass fiber	0.1	5/8	13	16	15	17	17	18	19
	0.6	2	14	17	18	21	25	29	35
Glass fiber with TL backing	0.6	2	14	17	18	22	27	32	39
Gypsum board tiles	1.8	1/2	14	16	18	18	21	22	22
Solid gypsum board ceiling	1.8	1/2	18	21	25	25	27	27	28
	2.3	5/8	20	23	27	27	29	29	30
Double layer of gypsum board	3.7	1	24	27	31	31	33	33	34
	4.5	1-1/4	26	29	33	33	35	35	36
Mineral fiber tiles, concealed spline mount	0.5 to 1	5/8	20	23	21	24	29	33	34

Example 8-2.

A 24 in. × 24 in. × 25 ft rectangular unlined supply duct is constructed of 22 gage sheet metal. Given the sound power levels in the duct (from row 1 of the table below), what are the breakout sound power levels 5 ft from the surface of the duct?

Solution.

Rows 1, 2, and 3 of the table below are self-explanatory, with rows 2 and 3 showing the calculations and rows 4 through 8 and 10 being calculated for 500 Hz as follows:

$$\gamma = 10^{(-\alpha/10)} = 10^{(-0.03/10)} = 0.99$$

$$L^* = \frac{\gamma^L - 1}{\ln\gamma} = \frac{0.99^{25} - 1}{\ln 0.99} = 23$$

$$S^* = (2)(L^*)\left(\frac{W+H}{12}\right) = (2)(23)\left(\frac{24+24}{12}\right) = 184$$

$$A = \frac{WH}{144} = \frac{(24)(24)}{144} = 4.0 \text{ ft}^2$$

Note that S/A only changed from 50 to 46 by using the more detailed method that considers duct attenuation. If the duct had been lined, the effect would have been more significant because there would have be more insertion loss (attenuation). The breakout sound power level from the duct surface is calculated with the following:

$$10\log_{10}(S^*/A) = 10\log_{10}(184/4) = 17 \text{ dB}$$

$$-10\log_{10}(\pi r L) + 10 = -10\log_{10}[\pi(5)(25)] + 10 = -16 \text{ dB}$$

Solution.

Row	Variable	Octave Band Center Frequency, Hz						
		63	125	250	500	1000	2000	4000
1	$L_{w(in)}$, dB	90	85	80	75	70	65	60
2	TL_{out}, dB (Table 8-6)	20	23	26	29	32	37	43
3	α, dB/ft (Table 8-10)	0.25	0.20	0.10	0.03	0.03	0.03	0.03
4	γ	0.94	0.95	0.98	0.99	0.99	0.99	0.99
5	L^*, ft	13	15	19	23	23	23	23
6	S^*, ft^2	106	119	152	184	184	184	184
7	A, ft^2	4.0	4.0	4.0	4.0	4.0	4.0	4.0
8	$10\log_{10}(S^*/A)$	14	15	16	17	17	17	17
9	$L_{w(out)}$, dB (Equation 8-5)	84	77	70	63	55	45	34
10	$-10\log_{10}(\pi r L)+10$	−16	−16	−16	−16	−16	−16	−16
11	L_p', dB (Equation 8-7)	68	61	54	47	39	29	18

Nonmetal Ducts

Whenever duct sound breakout is a concern, fiberglass or flexible round duct should not be used; these ducts have less transmission loss and are more transparent to sound. On the other hand, when there is a desire to have noise leave the duct system, these lower-mass systems can have advantages. For example, when noise from a horizontal outlet of a rooftop AHU (where the duct reenters the building) needs to be attenuated, some noise breaks out, reducing the amount of noise reentering the building.

SILENCERS

Duct silencers, also called *sound attenuators* or *sound traps*, reduce the noise transmitted from a source to the receiver (ASHRAE 2019). There are three types of silencers: active, dissipative or passive, and reactive.

Active Silencers

Active silencer systems are very effective at attenuating low-frequency, pure-tone noise in ducts (NEMI 2002). The rapid advancement of computer speeds allows fast Fourier transformation of sounds at higher frequencies. Alignment of sound in one direction (as in linear duct systems) is important for good cancellation. Active duct silencers consist of a microprocessor, two microphones placed a specific distance apart in a duct, and a speaker placed between the microphones (Figure 8-9). The microphone closest to the noise source senses the noise and the microprocessor converts it to an electrical signal then transmits a mirror-image sound wave to the speaker. When the waves collide in the duct, they essentially cancel each other out. The microphone downstream of the speaker senses the attenuated noise and sends a feedback signal to the microprocessor to adjust the speaker signal as necessary (NEMI 2002).

Active noise systems should be considered during the design phase, not only as a corrective action. One advantage of active silencers is that there is very little flow disruption (no pressure loss).

Dissipative Silencers

Dissipative silencers are by far the most common type of silencer. They can have rectangular or circular cross sections (Figure 8-10a) and are effective at attenuating broadband noise. However, they typically cause a pressure drop and a usually insignificant amount of generated noise in ducts; these elements should always be examined when dissipative silencers are considered (NEMI 2002).

Rectangular silencers usually are available in 3, 5, 7, and 10 ft lengths and several different configurations (Figure 8-11). Rectangular silencers have parallel sound-absorbing surfaces, usually perforated sheet metal, that cover cavities containing fiberglass or mineral wool (NEMI 2002).

Circular silencers are available in many open-face diameters, typically with lengths corresponding to the diameters. Most circular silencers have a center body similar to that shown in Figure 8-10b. The body is cylindrical, has a perforated sheet metal surface, and is filled with fiberglass or mineral wool. The outside shell may be of single-wall or double-wall construction. In single-wall construction, the shell is solid sheet metal with a diameter equal to the open-face diameter of the silencer. In double-wall construction, the shell is two concentric sheet metal shells, with the outer shell solid and the inner shell perforated and the space between the two filled with fiberglass or mineral wool; the diameter equal to the open-face diameter of the silencer (NEMI 2002). In addition to the perforated liner, the absorptive media may also include a tightly woven fiberglass cloth or a polymer film to further protect the insulation from erosion.

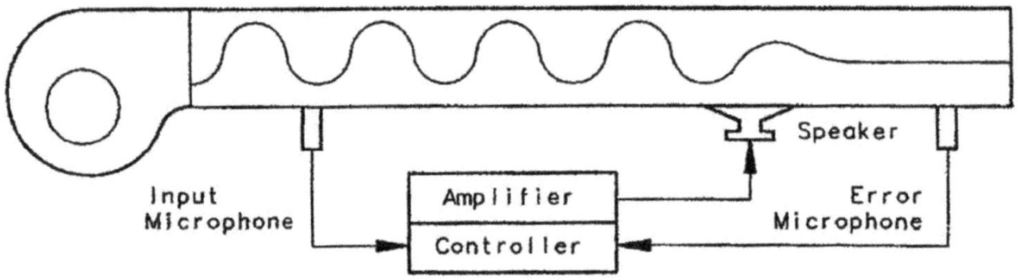

Figure 8-9 Active Duct Silencer
(Reprinted with permission from SMACNA 2004, Figure 5–13)

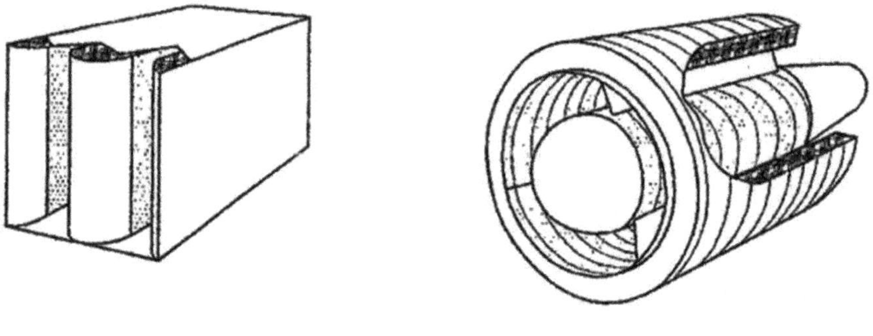

(a) Rectangular (b) Round

Figure 8-10 Dissipative Duct Silencers
(Reprinted with permission from SMACNA 2004, Figure 5–12)

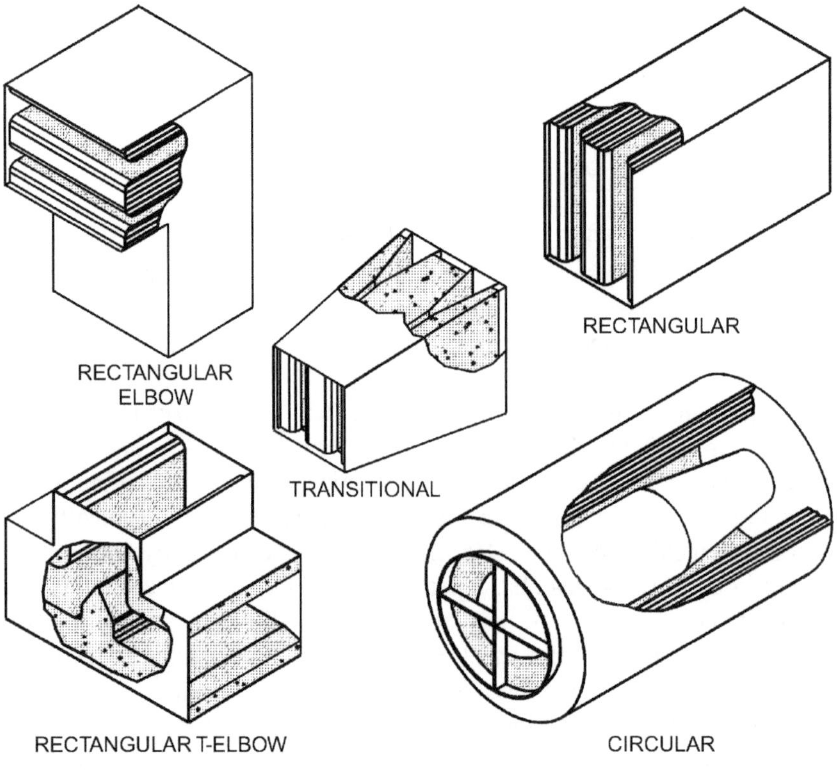

Figure 8-11 Duct Silencer Configurations
(Reprinted from ASHRAE 2019, Figure 21)

The insertion loss, generated noise, and pressure drop of a dissipative silencer are functions of the silencer's location in the duct system (supply or return) as well as its design (NEMI 2002). Dissipative silencers can have significant levels of insertion loss, but the insertion loss does not increase linearly with length. Most of the noise control is created in the first few feet of the silencer. Round silencers are available with an inner "torpedo" as shown in Figure 8-10b. No-loss round silencers are available without the center "torpedo." These typically use a perforated inner liner with fiberglass or mineral wool followed by a second solid liner, typically 2 to 8 in. larger in diameter. The advantages are that there is virtually no pressure drop and that any company with a spiral machine can make it with any thickness of insulation to meet low-frequency attenuation requirements. The disadvantage is less insertion loss per unit of length.

Reactive Silencers

Reactive silencers do not have absorption media. They are all metal, both solid and perforated, with chambers or baffles of special shapes that are tuned as resonators to react with and reduce sound power at certain frequencies (ASHRAE 2019). Without absorption media they usually have less noise reduction ability than dissipative silencers over a broad range of frequencies, although they can be extremely effective for narrow frequency ranges. In general, longer duct lengths and more pressure drop are needed for a given amount of sound attenuation. The main advantages of reactive silencers are that they may be washed down more readily and are often allowed in hospitals and biological laboratories. Also, there is no possibility of fiberglass entrainment because there is no fill.

SELECTING SILENCERS

Active and dissipative duct silencers complement each other. Active silencers are usually effective between the 16 and 250 Hz octave frequency bands, and dissipative silencers are effective from the 63 to 8000 Hz octave frequency bands (NEMI 2002). The general insertion losses, or attenuation characteristics, of active and dissipative duct silencers are shown in Figure 8-12. Within limits, dissipative silencers can be selected for better properties in the low or high frequencies. Longer dissipative silencers generally can provide a better ratio of insertion loss to pressure drop, but they cost more and occupy more space. Dissipative elbow silencers generally provide higher attenuation than straight silencers because there are no straight-through paths.

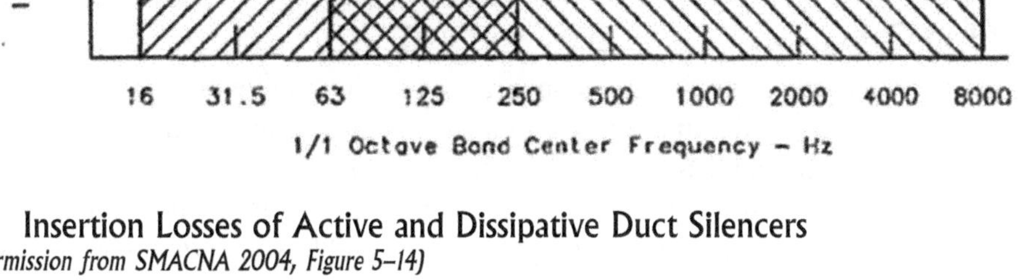

Figure 8-12 Insertion Losses of Active and Dissipative Duct Silencers
(Reprinted with permission from SMACNA 2004, Figure 5–14)

When selecting silencers, the designer has to be aware of not only the additional pressure loss that is introduced but also the generated noise level. Higher-pressure-loss silencers generate more noise, which is typically insignificant unless the silencer is near the air outlet. If after subtracting the insertion loss of the silencer its generated noise level is within 10 dB of the attenuated sound power level, the generated noise level must be added logarithmically to the attenuated sound power level to determines the actual sound power immediately downstream of the silencer (McGill 2003b).

DUCT SYSTEM ACOUSTICS

The main noise source in an HVAC system is generally the fan. Another source of noise is rapidly moving air; the noise it causes is referred to as *airflow-generated noise*. Duct designers should be concerned with two paths of noise propagation. Path 1 is the noise that propagates through the duct system, including noise generated at the fan and any noise generated along this path. Path 2 is the airborne noise radiated into surrounding spaces from the duct. Between these two propagation paths there is usually a trade-off. If a duct system is rigid and well-constructed, it will retain much of the noise within itself. This is desirable because for this propagation path there are several methods of noise attenuation available. Duct systems that are nonmetallic or made of lightweight materials radiate much of their in-duct noise to surrounding spaces. Although this may seem to produce significant in-duct attenuation, it can also be the source for radiated-noise problems in areas through which ducts pass (McGill 2003a).

This section examines 1) the primary noise source in HVAC systems—the fan, 2) the natural noise attenuation in a duct system, and 3) airflow-generated noise.

Fan Noise

Fan noise is a function of fan design, airflow rate, total pressure, and efficiency. Once a fan has been chosen, its size should be selected based on efficiency, because efficient operating ranges are also typically the quietest. Selecting size based on low outlet velocity alone is not appropriate, because low outlet velocity does not necessarily ensure quiet operation. Likewise, selections based on rotational or tip speed are not appropriate because noise comparisons of different fans (different types or the same type produced by different manufacturers) are not valid. Comparisons are only valid if they are of the actual sound power levels generated by the different types of fans producing the required total pressure and volume airflow rate (Bolton 1992).

Fan and air-handler manufacturers typically provide inlet, outlet, or both inlet and outlet sound power levels for their equipment. Most manufacturers provide test reports from independent laboratories rating the noise outputs of their products. It is very important to note the conditions under which the testing in these reports was performed since changes in airflow, total pressure, and point of operation (actual efficiency versus peak efficiency) can have a significant bearing on noise generation. Also, make sure the test data are for a product identical to that being considered (McGill 2003a). Fan manufacturers in some cases use the same data for inlet and outlet sound—the assumption is that the outlet sound is identical to the inlet sound, but that may not be the case. Today, most manufacturers separate inlet and outlet sound power levels, and both should be obtained if possible.

Point of Fan Operation

The point of fan operation affects acoustical output in a major way. It is common practice to select fans at their calculated point of maximum efficiency so that minimum power consumption is ensured. Generally, fan sound is at its minimum near the point of maximum efficiency for any given design. Figure 8-13 illustrates the location along the fan curve where fan noise is minimized. As the operating point shifts to the right (higher airflow and lower static pressure), noise increases.

At points to the left (lower airflow and higher static pressure), low-frequency noise can substantially increase; these operating points should therefore be avoided (ASHRAE 2019).

Blade-Pass Frequency

The blade-pass frequency is the number of times per second that the impeller of a fan passes the fan outlet. At this frequency all fans generate a tone and its multiples (harmonics). The tone being perceived as objectionable or as barely noticeable depends on the design of and type of fan as well as the point of operation (ASHRAE 2019).

System Effects

System effects can influence the sound power that a fan generates. These effects are pressure losses caused by non-uniform airflow at the inlet or discharge of a fan, which is caused by restrictions on inlets or outlets or other similar conditions (see Figure 8-14), for example, restricted inlets and inlet or outlet elbows being too close to the fan. Restricting inlet conditions can also increase sound levels at the blade-pass frequency (Greenheck. n.d.).

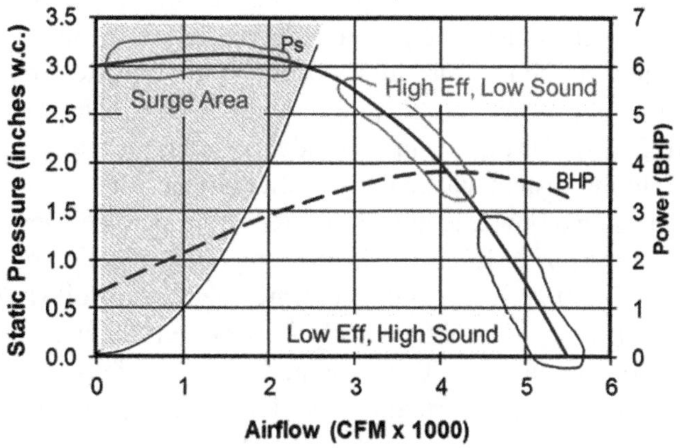

Figure 8-13 Fan Sound Levels
(Reprinted with permission from the course FAN-002, Energy Codes and Their Impact on Fan Selection, Greenheck Fan Corporation)

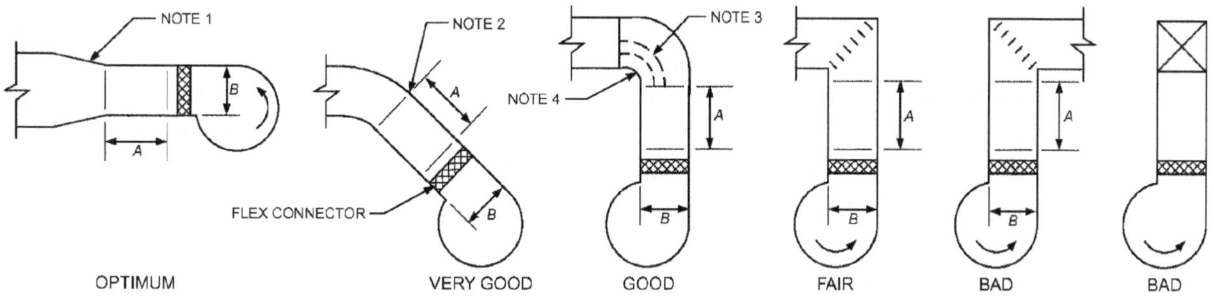

Notes:

1. Slopes of 1 in 7 preferred. Slopes of 1 in 4 permitted below 2000 fpm (10 m/s).
2. Dimension A should be at least 1.5 times B, where B is the largest discharge duct dimension.
3. Rugged turning vanes should be extended full radius of elbow.
4. Minimum 6 in. (150 mm) radius required.

Figure 8-14 Various Outlet Configurations for Centrifugal Fans and Their Possible Rumble Conditions
(Reprinted from ASHRAE 2019, Figure 25)

Plenum Fans

Plenum fans do not have housings around the fan impellers, and they discharge directly into the chamber, which pressurizes the plenum and forces air through ductwork. The air enters the fan impeller through an inlet bell at the chamber wall. If the plenum is sized appropriately and treated with sound-absorptive material, the fan can lower discharge sound power levels substantially.

The discharge from the plenum should not be near the fan's air blast to avoid increasing the blade-pass sound with the discharge air blowing directly into the duct. Inlets should not be obstructed, and the coils and filters should not be crowded (ASHRAE 2019). Chapter 49, "Noise and Vibration Control," of *ASHRAE Handbook—HVAC Applications* shows how to calculate attenuation due to plenums.

Minimizing Fan Sound

Proper fan selection and installation are vital to minimize the need for additional duct system sound attenuation (ASHRAE 2019). The following factors should be considered during fan selection:

- Design the air distribution system for minimum airflow resistance. If there is high system resistance then fans will operate at higher brake horsepower and therefore generate higher sound power levels (ASHRAE 2019). As a rule of thumb, increasing static pressure has twice as much effect on sound power as increasing flow rate.
- Carefully analyze the pressure losses of the system. System resistance that is higher than expected might produce sound power levels that are higher than had been estimated (ASHRAE 2019). During design, consider revising the fittings that have the highest pressure losses in the system, particularly in the critical path.
- Examine the sound power levels of different types and designs of fans. Fans that generate the lowest sound power levels (while meeting other selection requirements) should be chosen (ASHRAE 2019). In general, backward-inclined airfoil (BIAF) fans are quietest, followed by backward-inclined then lastly forward-curved fans. Mixed-flow fans can also be selected for quiet operation. Tubeaxial and vaneaxial fans can be quite loud. Manufacturers' ratings can provide information on a fan's sound power levels. The sound power produced depends on several factors other than the generic fan type.
- Consider additional acoustical treatment of the system if the fans generate objectionable tones and harmonics at the blade-pass frequency. The amplitudes of these tones can be affected by the duct system's resonance, by fan design, and by inlet flow distortions, which are created by poorly designed inlets or by inlet volume control damper operation (ASHRAE 2019). Today, use of variable-speed volume control instead of volume control dampers is strongly recommended.
- Design inlet and outlet duct connections for airflow that is straight and uniform. Avoid turbulent, unstable, or swirling inlet airflow (ASHRAE 2019). When fan noise problems do occur, they frequently exhibit significant energy in the 31.5 and 16 Hz octave bands (commonly called *rumble*). Refer to Figure 8-14 for guidance on inlet and outlet conditions.

SYSTEM ATTENUATION

Natural Attenuation

Attenuation is a reduction in noise level. Just as airflow loses energy as it propagates from the fan to the outlets, so does noise. Even without liners or silencers to control airborne noise, duct systems have mechanisms of natural attenuation of noise, such as elbow and end reflection, sound power splits, and duct wall losses (which are good in areas where the breakout noise is not an issue but breakout noises can be bad in other cases). Elbow reflection is the noise attenuation of sound being reflected by the any wall (of an elbow or other fitting) that changes the airflow direction within the duct. End reflection is the noise attenuation of sound being reflected back into the duct by any significant change in area at the end of a run of duct. Power splits are the natural sound attenuation that occurs when the sound energy is split by a divided-flow fitting. It is the ratio of one downstream path's cross-sectional area to the sum of all downstream paths' cross-sectional areas. All of these natural attenuation mechanisms together can provide significant noise reduction, and they may even eliminate the need for acoustically lined ductwork and/or silencers (McGill 2003a, 2003b).

Duct Attenuation

As sound pressure waves travel through a duct system, some portion of the sound pressure is transmitted to the surrounding surfaces. These surfaces vibrate slightly, dissipating some of the energy from the incident sound pressure wave (McGill 2003a). The vibrating duct surfaces also radiate sound to the surrounding surfaces, usually at a much lower level. This is called *breakout noise* and is covered in the earlier Breakout Noise subsection.

Duct attenuation for unlined sheet metal round and rectangular ducts is shown in Tables 8-9 and 8-10. Attenuation is expressed in terms of decibels per foot (dB/ft). For rectangular duct, natural attenuation is a function of the ratio of the duct's perimeter to the duct's cross-sectional area (P/A) for a 1 ft section. Noise attenuation is more significant in long lengths of duct. However, duct

Table 8-9 Sound Attenuation in Unlined Round Sheet Metal Ducts
(Reproduced from ASHRAE 2019, Table 19)

Duct Diameter, in.	Attenuation, dB/ft						
	Octave Band Center Frequency, Hz						
	63	125	250	500	1000	2000	4000
$D \le 7$	0.03	0.03	0.05	0.05	0.10	0.10	0.10
$7 < D \le 15$	0.03	0.03	0.03	0.05	0.07	0.07	0.07
$15 < D \le 30$	0.02	0.02	0.02	0.03	0.05	0.05	0.05
$30 < D \le 60$	0.01	0.01	0.01	0.02	0.02	0.02	0.02

Table 8-10 Sound Attenuation in Unlined Rectangular Sheet Metal Ducts
(Reproduced from ASHRAE 2019, Table 16)

Duct Size, in.	P/A, 1/ft	Insertion Loss, dB/ft						
		Octave Band Center Frequency, Hz						
		63	125	250	500	1000	2000	4000
6 × 6	8.0	0.30	0.20	0.10	0.10	0.10	0.10	0.10
12 × 12	4.0	0.35	0.20	0.10	0.06	0.06	0.06	0.06
12 × 24	3.0	0.40	0.20	0.10	0.05	0.05	0.05	0.05
24 × 24	2.0	0.25	0.20	0.10	0.03	0.03	0.03	0.03
48 × 48	1.0	0.15	0.10	0.07	0.02	0.02	0.02	0.02
72 × 72	0.7	0.10	0.10	0.05	0.02	0.02	0.02	0.02

Table 8-11 Insertion Losses of Unlined 90° Radius Rectangular Elbows
(Reproduced from ASHRAE 2019, Table 23)

Center Frequency × Width	Insertion Loss, dB
$fW < 1.9$	0
$1.9 \leq fW < 3.8$	1
$3.8 \leq fW < 7.5$	2
$fW > 7.5$	3

Note: $fW = f \times W$, where f = center frequency (kHz) and W = duct width (in.).

Table 8-12 Insertion Losses of Unlined and Lined Mitered Rectangular Elbows without Turning Vanes
(Reproduced from ASHRAE 2019, Table 22)

Center Frequency × Width	Insertion Loss, dB	
	Unlined Elbows	Lined Elbows
$fW < 1.9$	0	0
$1.9 \leq fW < 3.8$	1	1
$3.8 \leq fW < 7.5$	5	6
$7.5 \leq fW < 15$	8	11
$15 \leq fW \leq 30$	4	10
$fW > 30$	3	10

Note: $fW = f \times W$, where f = center frequency (kHz) and W = duct width (in.).

attenuation cannot reduce noise levels below the generated noise level of the air passing through the duct (McGill 2003a).

Especially in or near acoustically sensitive areas, caution must be exercised when nonmetallic ducts such as flexible duct, duct board, and fiberglass duct are considered. These ducts have very little mass and therefore have negligible transmission loss. A substantial amount of the noise inside such ducts radiates to surrounding surfaces, although some of these products do have high in-duct attenuation due to their absorptive wall surfaces (McGill 2003a).

Elbow Attenuation

When sound pressure waves traveling in a sheet metal duct encounter a hard surface such as the opposite wall of an elbow, some of the sound is reflected back in the direction of propagation (McGill 2003a). The result is a reduction in noise propagating past the elbow. Tables 8-11, 8-12, and 8-13 provide attenuation for rectangular radius and mitered unlined elbows with and without turning vanes as a function of elbow width for each octave band.

Branch Sound Power Splits

One of the most significant attenuation mechanisms is power splits. When sound traveling through a duct reaches a junction, the sound power in the incident sound in the main duct is distributed through the branches coming off the junction. The attenuation (reduction) of sound power that is transmitted down each branch off the junction has two components. The first is associated with

Table 8-13 Insertion Loss of Unlined and Lined Mitered Rectangular Elbows with Turning Vanes

(Reproduced from ASHRAE 2019, Table 24)

Center Frequency × Width	Insertion Loss, dB	
	Unlined Elbows	Lined Elbows
$fW < 1.9$	0	0
$1.9 \le fW < 3.8$	1	1
$3.8 \le fW < 7.5$	4	4
$7.5 \le fW < 15$	6	7
$fW > 15$	4	7

Note: $fW = f \times W$, where f = center frequency (kHz) and W = duct width (in.).

Table 8-14 Duct Branch Sound Power Division

(Reproduced from ASHRAE 2019, Table 26)

Area Ratio, $\frac{A_{bi}}{\Sigma A_{bi}}$	Sound Power Reduction (ΔL), dB	Area Ratio, $\frac{A_{bi}}{\Sigma A_{bi}}$	Sound Power Reduction (ΔL), dB
1.00	0	0.100	10
0.80	1	0.080	11
0.63	2	0.063	12
0.50	3	0.050	13
0.40	4	0.040	14
0.32	5	0.032	15
0.25	6	0.025	16
0.20	7	0.020	17
0.16	8	0.016	18
0.12	9	0.012	19

the reflection of the incident sound wave. The second and normally dominant component is associated with energy division according to the ratio of the cross-sectional area of each branch divided by the sum of the cross-sectional areas of the branches (NEMI 2002). The first component is neglected because it is less dominant and little data are available. The attenuation of sound power due to the second component is given by Equation 8-8 or Table 8-14. Sound level reductions due to power splits apply to all frequencies (McGill 2003a).

$$\Delta L_{bi} = 10\log_{10}\left(\frac{A_{bi}}{\Sigma A_{bi}}\right) \tag{8-8}$$

Example 8-3 shows how to determine sound attenuation for a branch fitting.

Example 8-3.
Determine the sound power attenuation for the straight-through branch (b1) and takeoff branch (b2) of a 12 × 12 × 6 in. tee. For rectangular duct assume the same vertical dimensions in all branches, 12 in. For round ducts the diameters can be squared and p can be ignored since it is in the numerator and denominator.

Solution.
The branch area ratios are 0.33 and 0.66 for rectangular and 0.20 and 0.80 for round, calculated as follows:

$$\text{Branch b1 for rectangular duct} = \frac{A_{b1}}{A_{b1} + A_{b2}} = \frac{6 \times 12}{(6 \times 12) + (12 \times 12)} = 0.33$$

$$\text{Branch b1 for round duct reduces to } \frac{D_1^2}{D_{b1}^2 + D_{b2}^2} = \frac{6^2}{6^2 + 12^2} = 0.20$$

$$\text{Branch b2 for rectangular duct} = \frac{A_{b2}}{A_{b1} + A_{b2}} = \frac{12 \times 12}{(6 \times 12) + (12 \times 12)} = 0.67$$

$$\text{Branch b2 for round duct reduces to } \frac{D_{b2}^2}{D_{b1}^2 + D_{b2}^2} = \frac{12^2}{6^2 + 12^2} = 0.80$$

From Table 8-14 the straight-through branch (b1) and takeoff branch (b2) attenuations are 5 and 2 dB for the rectangular duct and 7 and 1 dB for the round duct. Using Equation 8-8, the attenuation for the rectangular duct is

$$\Delta\text{dB b1} = 10\log_{10}(0.33) = -4.8 \text{ dB}$$

$$\Delta\text{dB b2} = 10\log_{10}(0.67) = -1.7 \text{ dB}$$

Attenuation for round duct is

$$\Delta\text{dB b1} = 10\log_{10}(0.20) = -7.0 \text{ dB}$$

$$\Delta\text{dB b2} = 10\log_{10}(0.80) = -1.0 \text{ dB}$$

Duct End Reflection Loss

When low-frequency sound waves reach a ductwork termination that is a large space, some of the sound energy is reflected back into the duct (ASHRAE 2019). Table 8-15 shows duct end reflection loss (ERL) values for duct terminating flush with a wall. To use Table 8-15 for rectangular duct, use Equation 8-9 to calculate the effective duct diameter:

$$D = \left(\frac{4A}{\pi}\right)^{1/2} \tag{8-9}$$

The duct ERL may be computed by Equation 8-10 for the frequency and duct sizes of interest (ASHRAE 2019):

$$\text{ERL} = 10\log_{10}\left[1 + \left(\frac{a_1 c_o}{\pi fD}\right)^{a_2}\right] \tag{8-10}$$

Table 8-15 Duct ERL: Duct Terminated Flush with Wall
(Reproduced from ASHRAE 2019, Table 28)

Duct Diameter, in.	ERL, dB						
	Octave Center Frequency, Hz						
	63	125	250	500	1000	2000	4000
6	18	12	7	3	1	0	0
8	15	10	5	2	1	0	0
10	14	8	4	1	0	0	0
12	12	7	3	1	0	0	0
16	10	5	2	1	0	0	0
20	8	4	1	0	0	0	0
24	7	3	1	0	0	0	0
28	6	2	1	0	0	0	0
32	5	2	1	0	0	0	0
36	4	2	0	0	0	0	0
48	3	1	0	0	0	0	0
72	1	0	0	0	0	0	0

Note: Table developed using Equation 8-10.

The quantities a_1 and a_2 are dimensionless constants determined as follows (ASHRAE 2019):

Termination	a_1	a_2
Flush	0.7	2
Free space	1	2

The ERL equation (Equation 8-10) has many limitations. For example, ducts with grille and blade-type diffuser terminations (even in suspended ceilings) are treated as having flush terminations, and free-space conditions require the outlet to be five duct diameters from any reflecting surface (such as a floor or a wall). Additionally, if flexible duct is included upstream of grilles, diffusers, or other terminal devices, the ERL will be reduced to near 0 above 63 Hz (ASHRAE 2019). Using Equation 8-9 to calculate D should yield reasonable results for diffusers and grilles with low aspect ratios (length/width).

Example 8-4 shows how to determine ERL.

AIRFLOW-GENERATED NOISE

Two main parameters determine the level of generated noise in a system: air velocity and turbulence. In HVAC systems, fans are a main source of sound, but aerodynamically generated sound may exceed fan sound due to the sound source's proximity to the sound receiver (e.g., when a variable-air-volume [VAV] damper near the diffuser is closed somewhat and the fan noise is substantially attenuated due to distance). Aerodynamically generated sound must be added logarithmically

Example 8-4.
Determine the duct ERL associated with a 12 in. round diffuser mounted flush in a suspended acoustical ceiling.

Solution.
Use Equation 8-10 for the calculations (this calculation should be done for each of the octave bands mid-frequency). The calculation for 250 Hz follows as an example:

$$\text{ERL} = 10\log_{10}\left[1+\left(\frac{(0.7)(1125)}{\pi(250)(12/12)}\right)\right] = 10\log_{10}[1+1.00^2] = 3.0$$

The table below summarizes the results for the remaining octave band center frequencies.

	Octave Band Center Frequency, Hz							
	63	125	250	500	1000	2000	4000	8000
$a_1 c_o/(\pi f D)$	3.98	2.01	.1.00	0.50	0.25	0.13	0.06	0.00
ERL, dB	12	7	3	1	0	0	0	0

in the element location's path when using source-path-receiver analysis to complete octave-band fan sound calculations (ASHRAE 2019).

Duct Velocities

In straight ducts with few or low pressure drop fittings, higher velocities may not be problematic in generating noise. Table 8-16 gives guidelines for recommended maximum airflow velocities in duct sections to avoid aerodynamically generated noise problems. In ducts, the amplitude of aerodynamically generated sound is roughly proportional to the fifth power of the duct airflow velocity in the vicinity of a duct element. Therefore, reducing velocity in a duct significantly reduces flow-generated noise (ASHRAE 2019).

Duct Fittings

Aerodynamically generated sound is created when turbulence occurs in the airflow at duct elements such as dampers, fittings, VAV boxes, silencers, and room air terminals. Generated noise increases as the pressure losses of ducts or fittings increase. Higher velocities or inefficient fittings cause higher levels of generated noise. To determine the airflow-generated noise of select fittings, consult *Algorithms for HVAC Acoustics* (Reynolds and Bledsoe 1991), which provides generated noise algorithms for dampers, elbows with turning vanes, junctions, and diffusers. If the airflow-generated noise of a fitting or any device is at least 10 dB below the section's sound power level, it will not contribute significantly to the overall sound power level (McGill 2003a).

When multiple duct fittings are installed close-coupled to each other, aerodynamic sound may increase because of the added turbulence (ASHRAE 2019). Layout is important. If the airflow does not become smooth and equalized before the next fitting or terminal device, the flow-generated noise in the next element will increase. It is best to maintain multiple diameters between duct elements. If duct fittings must be spaced closely, care should be exercised in designing them to minimize turbulence, flow separation, and distortion.

Table 8-16 Maximum Recommended Duct Airflow Velocities to Achieve Specified Acoustic Design Criteria
(Adapted from ASHRAE 2019, Table 8)

Duct Location	Design RC (N)	Maximum Airflow Velocity, fpm	
		Rectangular Duct	Round Duct
In shaft or above drywall ceiling	45	3500	5000
	35	2500	3500
	25	1700	2500
Above suspended acoustic ceiling	45	2500	4500
	35	1750	3000
	25	1200	2000
Within occupied space	45	2000	3900
	35	1450	2600
	25	950	1700

Table 8-17 Decibels to be Added to Diffuser Self-Generated Sound Rating to Allow for Throttling of Volume Damper
(Reproduced from ASHRAE 2019, Table 10)

Location of Volume Damper	Damper Pressure Ratio					
	1.5	2	2.5	3	4	6
	dB to be Added to Diffuser Sound Rating					
In neck of diffuser	5	9	12	15	18	24
In inlet of plenum of linear diffuser	2	3	4	5	6	9
In supply duct at least 5 ft from inlet plenum of linear diffuser	0	0	0	2	3	5

Volume Dampers

Volume dampers either radiate duct noise into occupied spaces through the ceiling or transmit it down the duct to room air outlets. When a manual damper is installed near an air terminal to balance the duct system, the air outlet's acoustical performance needs to be based on both the air volume handled and the magnitude of the air turbulence that the damper generates. The sound level that is produced when the damper is closed is accounted for in the air terminal sound rating by adding a correction factor. As the damper is modulated for air balance, this quantity is proportional to the throttled total pressure drop across the damper divided by the minimum total pressure drop across the damper, or the pressure ratio (ASHRAE 2019). Table 8-17 shows decibel corrections for determining the effect of damper location on diffuser self-generated noise rating.

Volume dampers used in sound-critical spaces should be placed several diameters from air terminals, with an acoustical lining included between the damper and the air terminal. Plenums that are acoustically lined can also be used between the room air terminal and the damper to reduce the damper sound (ASHRAE 2019).

Even in a properly designed and installed duct system, aerodynamically generated sound is affected by proper air balancing of the fan-duct system. Volume dampers in the critical path should always be wide open. If any damper in the critical path is even partially closed, that means the duct system is not balanced, and the fan will operate at a higher speed than necessary. The results are increases in fan power and fan noise and excessive aerodynamically generated sound in all duct elements (ASHRAE 2019).

Room Air Devices

Room air devices (grilles, diffusers, and registers) are rated for sound by manufacturers in terms of NC levels that typically include a receiver room sound correction of 10 dB. These NC ratings are useful for comparing different air devices against each other but should be added to other sound sources for determination of room sound levels (ASHRAE 2019).

To avoid problems associated with sound that is aerodynamically generated by air outlets, guidelines for recommended airflow velocities in duct outlets are provided in Table 8-18. Flexible connections between air devices and branch air ducts enable alignment of air devices with ceiling grids. Any misalignment in this connection, as shown in Figure 8-15, may cause up to 15 dB higher sound levels in the sound that is aerodynamically generated by the air device (ASHRAE 2019).

Avoiding Aerodynamically Generated Noise

Aerodynamically generated noise in a duct system can be reduced by the following actions:

- Sizing ducts for air velocities listed in Table 8-16 or lower.
- Avoiding sudden changes in direction or cross-sectional area (ASHRAE 2019).
- Providing smooth airflows at all ducts, branches (junctions), transitions, elbows, offsets, room air devices, and other duct elements (ASHRAE 2019).
- Providing straight ductwork between all duct elements (preferably 10 duct diameters or more apart) (ASHRAE 2019).

Table 8-18 Maximum Recommended Air Velocities at Neck of Supply Diffusers or Return Registers to Achieve Specified Acoustic Design Criteria
(Reproduced from ASHRAE 2019, Table 9)

Type of Opening	Design RC (N)	"Free" Opening Airflow Velocity, fpm
Supply air outlet	45	625
	40	560
	35	500
	30	425
	25	350
Return air opening	45	750
	40	675
	35	600
	30	500
	25	425

- Balancing duct systems for the lowest fan speed with dampers fully open in critical paths (ASHRAE 2019).
- Locating volume dampers at a minimum of 3 duct diameters from room air devices (preferably 5 to 10 duct diameters away) (ASHRAE 2019). Provide acoustically absorptive duct between the damper and the air device.
- Limiting flexible duct misalignment per Figure 8-15.

ROOM ACOUSTICS

At any given location, the sound pressure level of a particular sound source is a function of the sound power level and the acoustic properties of the room (its furnishings, its surface treatments, etc.), the room volume, the room absorption, the sound radiation characteristics of the sound source, and the distance between the sound source and the location (ASHRAE 2019). Refer to Chapter 49, "Noise and Vibration Control," of *ASHRAE Handbook—HVAC Applications* (2019) for detailed information about room acoustics.

Point Sound Sources

For point sound sources in relatively absorptive rooms (acoustical ceilings, commercial carpet, people), sound pressure levels decrease at a rate of roughly 3 dB per the doubled distance of the point of operation and the sound source (ASHRAE 2019). If the sound source is a single point source and the absorptive room has a volume less than 15,000 ft^3, Equation 8-11 can be used for converting sound power to pressure. For reflective rooms (e.g., with tile floors and/or gypsum

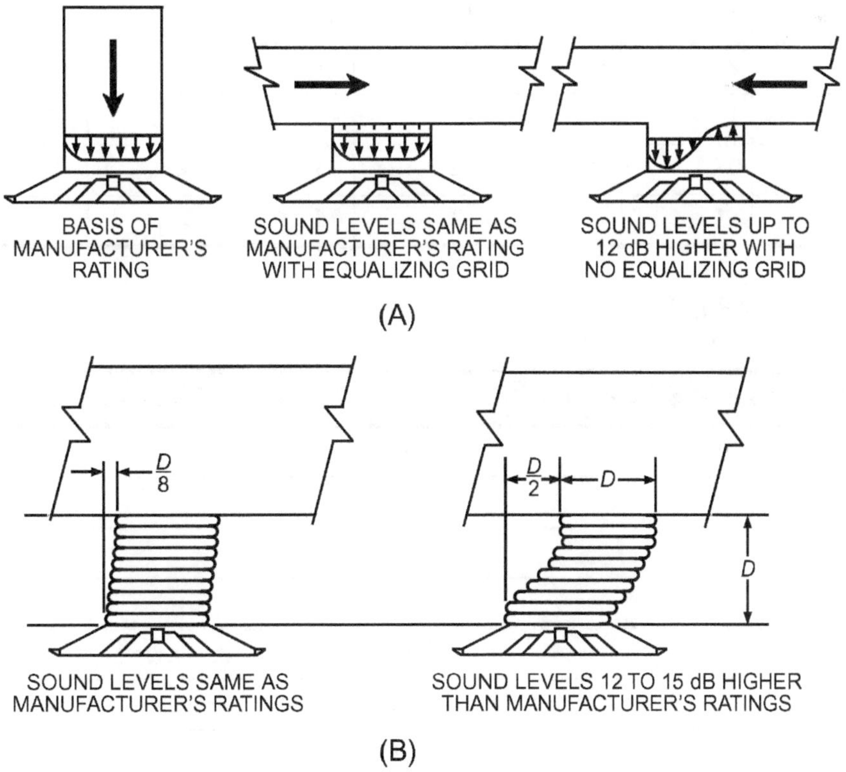

Figure 8-15 (a) Proper and Improper Air Conditions to an Outlet and (b) Effect of Proper and Improper Alignment of Flexible Duct to an Outlet
(Reprinted from ASHRAE 2019, Figure 14)

board walls and ceilings), the resulting sound pressure levels can differ from the calculations below.

$$L_p = L_w + A - B \tag{8-11}$$

where A is the value from Table 8-19 or Equation 8-12 and B is the value from Table 8-20 or Equation 8-13:

$$A = 5\log_{10}(V) - 3\log_{10}(f) + 25 \tag{8-12}$$

$$B = 10\log_{10}(r) \tag{8-13}$$

Example 8-5 shows how to determine octave band sound pressure level in a room where air is supplied by a single diffuser.

Table 8-19 Values for A in Equation 8-11
(Reproduced from ASHRAE 2019, Table 35)

Room Volume (V), ft^3	Values for A, dB						
	Octave Band Center Frequency (f), Hz						
	63	125	250	500	1000	2000	4000
1500	4	3	2	1	0	−1	−2
2500	3	2	1	0	−1	−2	−3
4000	2	1	0	−1	−2	−3	−4
6000	1	0	−1	−2	−3	−4	−5
10,000	0	−1	−2	−3	−4	−5	−6
15,000	−1	−2	−3	−4	−5	−6	−7

Table 8-20 Values for B in Equation 8-11
(Reproduced from ASHRAE 2019, Table 36)

Distance from Sound Source (r), ft	Values for B, dB
3	5
4	6
5	7
6	8
8	9
10	10
13	11
16	12
20	13

More Complex Rooms

Many rooms have multiple diffusers or tile floors or other features that make them more complex when it comes to determining their sound pressure levels. Chapter 49, "Noise and Vibration Control," of *ASHRAE Handbook—HVAC Applications* (2019) offers information on determining sound pressure levels for more complex rooms.

For example, in many conditioned spaces, the air supply outlets are located flush with the ceiling of the space and effectively serve as an array of distributed ceiling sound sources (ASHRAE 2019).

Nonstandard rooms are also more complex. Equations 8-11, 8-12, and 8-13 assume that the subject rooms are normally furnished. Do not use these equations for spaces that have little sound absorption, e.g., sports or athletic areas, concert halls, electrical and some mechanical rooms, and operating rooms. The previous analysis can underestimate the actual sound power levels by as much as 10 to 15 dB when the room is reflective.

ROOM DESIGN GUIDELINES

The design guidelines for HVAC-related background sound for various types of space occupancies are given in Table 8-21. When sound from non-HVAC sources, such as traffic and office equipment, are considered, use the lower sound level as the basis for design. The design goal is to attain a balanced sound spectrum with no single frequency range dominant.

Example 8-5.
A 14 × 20 × 8 ft office is supplied by one diffuser. The diffuser is mounted on the longitudinal centerline of the 14 ft wide room, 5 ft from one end wall. The sound power level upstream of the diffuser and the generated noise level (GNL) of the diffuser are shown in Table 8-17. What are the octave band sound pressure levels at a height of 5 ft from the floor, on the longitudinal centerline, 10 ft from the same end of the room?

Solution.
Rows 1 and 2 of the table below are inputs and row 3 is obtained by logarithmic addition of rows 1 and 2. The room volume is 2240 ft^3. Using trigonometry, the distance from the diffuser is 5.8 ft, calculated as follows:

$$\sqrt{(8-5)^2 + (10-5)^2} = 5.83 \text{ ft}$$

The sound pressure levels 5.8 ft from the sound source, calculated by Equation 8-12, are summarized in Table 8-15.

Row	Frequency, Hz	63	125	250	500	1000	2000	4000
1	Duct sound power L_w, dB	72	73	63	55	49	46	42
2	Diffuser GNL, dB	45	52	59	54	42	38	29
3	Room sound power L_w, dB	72	73	65	58	50	47	42
4	A from Table 8-13	3	2	1	0	−1	−2	−3
5	B from Table 8-14	8	8	8	8	8	8	8
6	Room sound pressure L_p, dB (Equation 8-11)	67	67	58	50	41	37	31

Table 8-21 Design Guidelines for HVAC-Related Background Sound in Rooms
(Reproduced from ASHRAE 2019, Table 1)

Room Type	RC (N)	Room Type	RC (N)
Residences, Apartments, Condominiums	30	Hospitals and Clinics	
Hotels/Motels		Private rooms	30
Individual rooms or suites	25–35	Operating rooms	35
Meeting/banquet rooms	25–35	Corridors and public areas	40
Corridors, lobbies	35–45	Laboratories (with fume hoods)	
Service/support areas	35–45	Testing/research, minimal speech communication	50
Office Buildings		Research, extensive telephone use, speech communication	40–50
Executive and private offices	25–35	Group teaching	35–45
Conference rooms	25–35	Churches, Mosques, Synagogues	
Teleconference rooms	25 max	General assembly with critical music programs*	25–35
Open-plan offices	40 max	Schools	
With sound masking	35 max	Classrooms	25 to 30
Circulation and lobbies	40–45	Large lecture rooms	25 to 30
Performing Arts Spaces*		Without speech amplification	25 max
Drama theaters, concert and recital halls	25	Libraries	30–40
Music teaching studios	25	Indoor Stadiums and Gymnasiums	
Music practice rooms	30–35	Gymnasiums and natatoriums	40–50
Courtrooms		Large-seating-capacity spaces with amplified speech	45–55
Unamplified speech	25–35		
Amplified speech	30–40		

*An experienced acoustical consultant should be retained for guidance on acoustically critical spaces (below RC 30) and for all performing arts spaces.

SUPPLEMENTAL ATTENUATION REQUIREMENTS

When more noise control is required than the natural attenuation of the duct system provides, designers can select from various methods. One obvious method is simply to use a quieter fan. Another method is to line ductwork with insulation—typically fiberglass, but foamed rubber products are available and effective in especially the 500 and 1000 Hz octave bands. Designers may also choose to add plenum or duct silencers to further attenuate noise. Near fans these are normally selected for 125 and 250 Hz attenuation, and downstream of VAV boxes they are usually selected for higher-frequency attenuation. A combination of any or all three of these methods may be necessary, depending on the noise criteria established by the owner and the cost to implement. It is important to know the noise criteria by octave, because the method of treatment will vary depending on whether the problem is in lower, middle, or higher octave bands or a combination of these.

Adding liners to the interior of ductwork can significantly increase noise attenuation. There are acoustical benefits to lined rectangular, round, and flat oval duct, as well as use of double wall duct liner where the inner wall is perforated to allow sound absorption by fiberglass between the perforations and the exterior wall of the duct. Using insulation in elbows increases the sound absorption. Designers should be careful when considering double-wall duct or exterior insulation wrapping applications. Using a solid inner liner or lagging the ductwork on the outside does not have a significant effect on the control of noise inside the duct that reaches outlets, although lagging is effective in reducing breakout noise. If insulation is necessary to control heat transfer or prevent condensation, properly applied internal insulation can also provide noise control. Interior insulation, whether lined or perforated inner double-wall duct, increases the system total pressure drop slightly.

Acoustically Lined Round Sheet Metal Ducts

Tables 8-22 and 8-23 show insertion loss values in decibels for round double-wall sheet metal ducts with 1 and 2 in. acoustical lining, respectively, for 5, 10, 20, and 40 ft lengths (ASHRAE 2019). The acoustical lining on the ducts researched to obtain these values consists of a fiberglass blanket with an internal liner that is perforated sheet metal having an open area of 23%. There are many available material options for round ducts, so the data provided in these tables may differ significantly from real-world attenuation values (ASHRAE 2015).

Acoustically Lined Rectangular Sheet Metal Ducts

Internal duct linings can provide sound attenuation as well as thermal insulation for rectangular sheet metal ducts (ASHRAE 2019). It is important that the vapor barrier is in the correct location when using liners. In cold climates, outdoor ducts carrying humidified air should not be lined. Duct lining thickness for thermal insulation usually varies from 0.5 to 2 in. (NEMI 2002). Per ASTM C1071 (2019), the density of fiberglass lining is no longer required to be reported, as it is not a good indicator of acoustical or thermal performance. Figures 8-16 and 8-17 provide attenuation values of selected rectangular sheet metal duct sizes for 1 and 2 in. fiberglass lining (Reynolds et al. 2018).

Acoustically Lined Rectangular Elbows

Table 8-12 gives insertion loss values for unlined and lined mitered rectangular elbows without turning vanes and Table 8-13 gives the values for these elbows with turning vanes. For lined elbows, the duct lining must extend at least two duct widths beyond the elbow as originally tested (NEMI 2002). These tables apply only where the duct is lined before and after the elbow. No data are available for round or radius rectangular elbows.

It is interesting to note that lined mitered elbows without turning vanes provide more attenuation than lined mitered elbows with turning vanes. Turning vanes help the airflow through the elbow, but they also deflect sound waves through the elbow, preventing total exposure of the acoustical energy to the lining on the surfaces of the elbow. Acoustical turning vanes are available, but the acoustical attenuation is not large.

COMPUTER ANALYSIS

The use of an acoustics program is highly recommended to dramatically shorten the design time needed to accurately calculate the acoustics equations. Acoustic estimates should be part of every project where the designer's experience indicates that there could be issues—not just projects that can afford acoustical consultants. After a building is built, sound-reducing fixes tend to be expensive because physical constraints typically eliminate many of the simpler, lower-cost options. After installation, changing the fan type or size is not feasible. Instead, the solution is possibly to add lead, gypsum board, or other materials that have high transmission losses.

Table 8-22 Insertion Losses for Round Sheet Metal Ducts with 1 in. Thick Fiberglass Lining
(Reproduced from ASHRAE 2019, Table 20)

Diameter, in. ID	Length, ft	Insertion Loss, dB Octave Midband Frequency, Hz							
		63	125	250	500	1000	2000	4000	8000
6	10	4	6	9	12	22	23	20	13
8	10	3	5	9	15	22	22	18	12
10	10	2	4	8	15	22	20	16	12
12	5	0	1	4	12	28	29	16	14
	10	0	3	7	19	41	35	21	18
	20	0	5	13	33	50	48	30	26
	40	0	9	24	50	50	50	44	39
16	5	0	1	3	11	26	22	16	16
	10	0	2	6	17	38	27	19	18
	20	0	4	11	29	50	37	25	23
	40	0	7	20	49	50	50	36	32
20	5	0	1	3	10	23	17	14	16
	10	0	2	5	15	33	21	17	17
	20	0	3	9	25	48	28	21	20
	40	0	6	16	44	50	42	29	25
24	5	0	1	2	9	20	13	12	14
	10	0	1	4	14	29	16	14	15
	20	0	2	7	22	44	22	17	17
	40	0	4	13	39	50	32	23	21
28	5	0	1	2	8	17	10	11	12
	10	0	1	3	12	24	12	12	13
	20	0	2	6	20	38	17	14	15
	40	0	4	11	34	50	25	19	18
32	5	0	1	2	8	14	8	9	10
	10	0	1	3	11	20	10	10	11
	20	0	2	5	18	31	14	12	13
	40	0	3	10	31	49	21	17	17
36	5	0	1	1	7	13	7	7	8
	10	0	1	3	11	17	9	8	9
	20	0	1	5	17	25	12	11	11
	40	0	3	9	29	41	18	15	16
40	5	0	0	1	7	12	7	6	6
	10	0	1	3	10	15	8	7	7
	20	0	1	5	17	20	11	10	10
	40	0	3	9	27	31	16	15	15
44	5	0	0	1	7	11	7	6	5
	10	0	1	2	10	13	8	7	6
	20	0	1	5	17	17	10	9	9
	40	0	3	8	27	23	14	13	14
48	5	0	0	1	7	12	6	6	5
	10	0	1	2	10	13	7	7	6
	20	0	1	4	17	16	9	8	8
	40	0	3	8	29	20	13	11	13
54	10	0	0	1	4	1	2	3	3
60	10	0	0	0	0	1	1	1	1

Table 8-23 Insertion Losses for Round Sheet Metal Ducts with 2 in. Thick Fiberglass Lining
(Reprinted from ASHRAE 2019, Table 21)

Diameter, in. ID	Length, ft	Insertion Loss, dB Octave Midband Frequency, Hz							
		63	125	250	500	1000	2000	4000	8000
6	10	6	8	14	22	22	23	20	13
8	10	5	7	13	22	22	22	18	12
10	10	4	7	13	22	22	20	16	11
12	5	2	4	10	25	41	29	18	16
	10	3	7	17	35	46	37	23	18
	20	5	13	31	50	50	49	32	22
	40	7	22	50	50	50	50	47	29
16	5	2	3	8	24	35	21	17	17
	10	3	6	14	34	42	28	21	19
	20	4	10	25	50	50	39	27	21
	40	5	18	46	50	50	50	39	25
20	5	2	3	6	23	27	16	15	17
	10	2	5	12	32	35	21	18	18
	20	3	8	21	48	48	29	23	20
	40	4	15	39	50	50	43	31	22
24	5	1	2	5	22	21	13	14	16
	10	2	4	10	30	29	16	15	16
	20	2	7	18	45	43	21	18	18
	40	3	12	33	50	50	31	24	20
28	5	1	2	5	21	17	10	12	13
	10	1	3	8	29	24	12	13	14
	20	2	6	15	43	37	16	15	15
	40	3	11	28	50	50	23	19	18
32	5	1	1	4	20	15	9	11	11
	10	1	3	8	27	21	11	12	12
	20	2	5	14	42	31	13	13	13
	40	2	10	25	50	49	18	17	17
36	5	1	1	4	19	14	8	9	8
	10	1	2	7	26	18	10	10	9
	20	1	5	13	41	25	12	12	11
	40	3	9	23	50	40	17	15	16
40	5	1	1	4	19	12	8	8	6
	10	1	2	7	26	15	9	9	7
	20	2	4	12	40	20	12	11	10
	40	3	9	22	50	29	16	15	15
44	5	1	1	4	19	11	7	6	4
	10	1	2	7	26	13	8	7	6
	20	2	4	12	39	16	11	10	9
	40	3	8	21	50	22	16	16	16
48	5	1	1	4	18	10	5	5	4
	10	1	2	7	25	12	7	6	6
	20	2	4	12	38	15	9	9	10
	40	3	7	20	50	21	15	16	17
54	10	0	1	6	11	1	2	3	2
60	10	0	0	5	8	1	1	0	0

		1/1 Octave Band Center Frequency							
L (ft)	P/A (1/in)	63	125	250	500	1000	2000	4000	8000
5	0.278	2	1	3	12	28	23	16	13
5	0.229	1	1	3	11	25	18	12	10
5	0.180	1	1	2	9	21	14	9	8
5	0.131	0	0	2	8	15	10	6	5
5	0.082	0	0	1	6	9	6	4	3
10	0.278	2	2	6	18	42	34	20	16
10	0.229	1	1	5	17	41	26	16	14
10	0.180	1	1	4	15	36	19	12	11
10	0.131	1	1	3	13	26	13	9	7
10	0.082	0	1	3	10	11	7	5	4
15	0.278	1	2	9	24	50	44	25	19
15	0.229	1	2	7	22	50	33	20	17
15	0.180	1	2	6	20	49	24	16	14
15	0.131	1	1	5	17	35	15	11	9
15	0.082	0	1	4	14	13	9	6	4
20	0.278	1	3	11	30	50	50	29	22
20	0.229	1	2	10	28	50	40	24	20
20	0.180	1	2	8	25	50	28	19	16
20	0.131	1	2	6	22	45	18	13	12
20	0.082	0	1	5	17	15	10	7	5
25	0.278	2	3	14	36	50	50	33	26
25	0.229	1	3	12	33	50	46	28	23
25	0.180	1	3	10	30	50	32	22	19
25	0.131	1	2	8	26	50	21	16	14
25	0.082	0	1	6	20	17	12	8	6
30	0.278	2	4	17	41	50	50	37	29
30	0.229	1	3	14	39	50	50	32	27
30	0.180	1	3	11	35	50	37	26	22
30	0.131	1	3	9	30	50	23	18	16
30	0.082	0	2	7	23	19	13	9	7
40	0.278	2	4	22	49	50	50	44	35
40	0.229	1	4	18	48	50	50	40	33
40	0.180	2	4	15	44	50	45	32	28
40	0.131	2	3	12	38	50	28	23	20
40	0.082	0	2	9	29	23	16	12	9

Figure 8-16 Insertion Losses for Rectangular Sheet Metal Ducts with 1 in. Thick Fiberglass Lining
(Reprinted from Reynolds et al. 2018, Table 5-13)

NOMENCLATURE

a = dimensionless coefficient
A = cross-sectional area of duct, ft^2
A_{bi} = cross-sectional area of branch i, in.^{2}
B = $10\log_{10}(r)$
c_o = speed of sound in air, 1125 fps at standard conditions
D = duct diameter, ft
ERL = duct-end acoustical power reflection loss, dB
f = frequency, Hz
H = rectangular duct height, in.
L = length, ft
L^* = effective duct length, ft
L_p = sound pressure level at a specified distance from sound source, dB (re 20 μPa)
L_p' = estimate of breakout sound pressure level at a specific point in space in rooms with exposed ductwork, dB (re 10^{-12} W)

L (ft)	P/A (1/in)	1/1 Octave Band Center Frequency 63	125	250	500	1000	2000	4000	8000
5	0.278	1	3	9	22	40	19	16	13
5	0.229	1	3	8	21	33	15	12	11
5	0.180	1	2	6	19	23	11	9	8
5	0.131	1	1	6	16	15	8	6	5
5	0.082	0	1	5	12	8	5	4	3
10	0.278	1	4	16	38	50	28	20	17
10	0.229	2	4	14	35	50	22	17	15
10	0.180	1	4	12	31	36	17	13	12
10	0.131	1	3	10	27	21	11	9	8
10	0.082	0	2	8	21	10	7	5	4
15	0.278	1	4	22	50	50	37	25	21
15	0.229	2	5	20	48	50	29	21	19
15	0.180	1	5	17	44	46	22	17	16
15	0.131	1	4	14	38	27	14	12	11
15	0.082	0	3	11	29	11	8	6	5
20	0.278	2	4	28	50	50	46	29	25
20	0.229	2	6	25	50	50	36	26	23
20	0.180	2	6	22	50	50	27	21	20
20	0.131	1	5	18	48	31	17	15	14
20	0.082	1	3	14	36	12	9	7	6
25	0.278	2	5	34	50	50	50	34	28
25	0.229	2	6	30	50	50	43	30	28
25	0.180	2	7	27	50	50	31	25	24
25	0.131	1	6	22	50	36	20	17	17
25	0.082	1	4	17	43	13	10	8	7
30	0.278	2	5	40	50	50	50	38	32
30	0.229	2	7	36	50	50	48	35	32
30	0.180	2	8	31	50	50	36	29	28
30	0.131	2	7	26	50	41	23	20	20
30	0.082	1	5	20	48	14	11	9	8
40	0.278	2	5	48	50	50	50	45	39
40	0.229	2	8	45	50	50	50	44	40
40	0.180	2	10	40	50	50	45	37	35
40	0.131	2	10	33	50	48	29	25	26
40	0.082	1	6	25	50	15	14	11	11

Figure 8-17 Insertion Losses for Rectangular Sheet Metal Ducts with 2 in. Thick Fiberglass Lining
(Reprinted from Reynolds et al. 2018, Table 5–14)

L_w = sound power level of sound source, dB (re 10^{-12} W)
$L_{w(in)}$ = sound power level of sound inside duct at location where breakout noise is of concern, dB (re 10^{-12} W)
$L_{w(out)}$ = sound power level of sound radiated from outside surface of duct walls, dB (re 10^{-12} W)
P = acoustic pressure, N/m^2 or Pa
P_{ref} = standard reference sound pressure, 2×10^{-5} Pa (threshold of excellent youthful hearing)
r = distance between sound source and receiver, ft
S = surface area of outside sound-radiating surface of duct, ft^2
S^* = effective surface area of duct, ft^2
TL_{out} = normalized (independent of S and A) duct breakout transmission loss, dB
V = room volume, ft^3
w = sound power, W
W = rectangular duct width, in.
w_{ref} = reference sound power = 10^{-12} W

Symbols

α = duct attenuation rate, dB/ft

ΔL_{bi} = attenuation of sound power at a junction that is related to the sound power transmitted down an individual branch of the junction (sound power reduction), dB

γ = coefficient

ΣA_{bi} = total cross-sectional area of the individual branches, in.^{2}

REFERENCES

ASHRAE. 2015. Chapter 48, Noise and vibration control. In *ASHRAE handbook—HVAC applications*. Peachtree Corners, GA: ASHRAE.

ASHRAE. 2017. Chapter 8, Sound and vibration. In *ASHRAE handbook—Fundamentals*. Peachtree Corners, GA: ASHRAE.

ASHRAE. 2019. Chapter 49, Noise and vibration control. In *ASHRAE handbook—HVAC applications*. Peachtree Corners, GA: ASHRAE.

ASTM. 2019. ASTM C1071-19, *Standard specification for fibrous glass duct lining insulation (thermal and sound absorbing material)*. West Conshohocken, PA: ASTM International.

Beranek, L.L. 1989. Balanced noise criteria (NCB) curves. *Journal of the Acoustic Society of America* 86:650–54.

Bolton, A.N. 1992. *Intercomparison of in-duct fan noise measurements*. Report EUR 13890 EN. Brussels, Belgium: Commission of the European Communities, Community Bureau of Reference.

Greenheck. n.d. Centrifugal fan catalogs. Schofield, WI: Greenheck Fan Corporation.

McGill. 2003a. Chapter 8, Duct system acoustics. In *Duct System Design Guide*. Groveport, OH: McGill AirFlow Corporation.

McGill. 2003b. Chapter 10, Supplemental attenuation. In *Duct System Design Guide*. Groveport, OH: McGill AirFlow Corporation.

NEMI. 2002. Appendix F, Acoustics. In *Final strategic planning report*. Alexandria, VA: National Energy Management Institute, National Center for Energy Management and Building Technologies.

OSHA. n.d. Occupational noise exposure. *Code of federal regulations*. 29 CFR 1910.95. www.osha.gov/laws-regs/regulations/standardnumber/1910/1910.95.

Reynolds, D.D., and J.M. Bledsoe. 1991. *Algorithms for HVAC acoustics*. Peachtree Corners, GA: ASHRAE.

Reynolds, D., M.A. Schwob, and J. Albright. 2018. *The effect of lining length on the insertion loss of acoustical duct liner in sheet metal ductwork*. Final report of ASHRAE Research Project RP-1408. Peachtree Corners, GA: ASHRAE.

Schaffer, M.E. 1991. *A practical guide to noise and vibration control for HVAC systems*. Peachtree Corners, GA: ASHRAE.

Schaffer, M.E. 2005. *A practical guide to noise and vibration control for HVAC systems*, 2nd ed. (I-P). Peachtree Corners, GA: ASHRAE.

SMACNA. 2004. *HVAC sound and vibration manual*. SMACNA 1858-2004. Chantilly, VA: Sheet Metal and Air Conditioning Contractors' National Association.

BIBLIOGRAPHY

Simmons, R. 2997. Vibration isolation. *ASHRAE Journal* 49(8):30–40.

Tinsley, R.W. 2002. Classroom noise control. *HPAC Engineering* 71(10):41–46.

Yerges, J.F., and J.R. Yerges. 1997. Noise and vibration control for HVAC and piping systems. *Heating/Piping/Air Conditioning* 69(10):44–47.